SOLUTIONS MANUAL

ADVANCED ORGANIC CHEMISTRY

BERNARD MILLER
University of Massachusetts, Amherst

Upper Saddle River, NJ 07458

Project Manager: Kristen Kaiser
Editor-in-Chief, Science: John Challice
Vice President of Production & Manufacturing: David W. Riccardi
Executive Managing Editor: Kathleen Schiaparelli
Assistant Managing Editor: Becca Richter
Production Editor: Rhonda Aversa
Supplement Cover Manager: Paul Gourhan
Supplement Cover Designer: Joanne Alexandris
Manufacturing Buyer: Ilene Kahn
Cover Image Credit: Ryan McVay/Getty Images, Inc.

Pearson Prentice Hall
Pearson Education, Inc.
Upper Saddle River, NJ 07458

Printed in the United States of America

10 9 8 7 6 5 4 3 2 1

ISBN 0-13-101443-9

Pearson Education Ltd., *London*
Pearson Education Australia Pty. Ltd., *Sydney*
Pearson Education Singapore, Pte. Ltd.
Pearson Education North Asia Ltd., *Hong Kong*
Pearson Education Canada, Inc., *Toronto*
Pearson Educación de Mexico, S.A. de C.V.
Pearson Education—Japan, *Tokyo*
Pearson Education Malaysia, Pte. Ltd.

Table of Contents

CHAPTER 1

INTRODUCTION

PROBLEM SOLUTIONS

1a. (3-methylcyclohexene) + $H\text{-}OSO_3H$ ⟶ (carbocation) H $^{\ominus}OSO_3H$ ⟶ (1-methylcyclohexene) + H_2SO_4

1b. $CH_3CHBrCH{=}CH_2$ ⟶ $[CH_3\overset{+}{C}H{-}CH{=}CH_2 \longleftrightarrow CH_3CH{=}CH{-}\overset{+}{C}H_2]\ Br^{\ominus}$

$CH_3CH{=}CH{-}\overset{+}{C}H_2$ $Br^{\ominus}$ $HO{-}CH_3$ ⟶ $CH_3CH{=}CH{-}CH_2{-}\overset{+}{O}(H){-}CH_3$ $Br^{\ominus}$

↓

$CH_3CH{=}CH{-}CH_2{-}OCH_3$ + HBr

or:

$CH_3\overset{+}{C}H{-}CH{=}CH_2$ $HO{-}CH_3$ ⟶ $CH_3CH{=}CH{-}CH_2{-}\overset{+}{O}(H){-}CH_3$ $Br^{\ominus}$

↓

$CH_3CH{=}CH{-}CH_2{-}OCH_3$ + HBr

As always, it is simply a matter of convenience which resonance form is used, since the different resonance forms represent the same ion, and must yield the same products.

1c. 1-methylcyclopentene + H–Cl → (cation, H, $Cl^{\ominus}$) → 1-methylcyclopentene + H–Cl → tertiary cation + $Cl^{\ominus}$ → 1-chloro-1-methylcyclopentane

CH_3 | H_3C H $Cl^{\ominus}$ ⊕ | CH_3 H–Cl | CH_3 $^{\ominus}Cl$ ⊕ | H_3C Cl

1d. HO_3SO–H

$H_2C{=}CH{-}CH{=}CH{-}CH_3 \longrightarrow H_3C{-}CH{=}CH{-}\overset{+}{C}H{-}CH_3 + HSO_4^{\ominus}$ (OH_2)

$\longrightarrow H_3C{-}CH{=}CH{-}CH(\overset{+}{O}H_2){-}CH_3$ ($^{\ominus}OSO_3H$) $\longrightarrow H_3C{-}CH{=}CH{-}CH(OH){-}CH_3 + H_2SO_4$

1e. Cl–H

$CH_3{-}CH{=}CH{-}OCH_3 \longrightarrow CH_3{-}CH_2{-}CH{=}\overset{+}{O}CH_3$ ($Cl^{\ominus}$, OH_2) $\longrightarrow CH_3{-}CH_2{-}CH(\overset{+}{O}H_2){-}OCH_3$ ($Cl^{\ominus}$)

$\longrightarrow CH_3{-}CH_2{-}CH(OH){-}OCH_3$ + H–Cl $\longrightarrow CH_3{-}CH_2{-}CH{-}CH(HO){-}\overset{+}{O}CH_3(H)$ $Cl^{\ominus}$

$\longrightarrow CH_3OH + CH_3{-}CH_2{-}CH{=}\overset{+}{O}{-}H$ ($Cl^{\ominus}$) $\longrightarrow CH_3{-}CH_2{-}CH{=}O + HCl + HOCH_3$

An incorrect mechanism

You may have proposed the mechanism shown below. That would not be a very unreasonable suggestion, but has to be rejected because forming the vinyl carbocation requires pulling an electron pair out of an sp^2 orbital. As a result, vinyl carbocations are very high energy intermediates, which are almost never formed except when they are stabilized by strongly electron-donating substituents.

$CH_3CH{=}CH{-}OCH_3 + H{-}Cl \longrightarrow CH_3CH{=}CH{-}\overset{+}{O}(H)CH_3 \xrightarrow{??} CH_3CH{=}\overset{+}{C}H + HOCH_3$

$CH_3CH{=}\overset{+}{C}H + OH_2 \longrightarrow CH_3CH{=}CH{-}\overset{+}{O}H_2 \xrightarrow{Cl^-} CH_3CH{=}CH{-}OH \xrightarrow{H{-}Cl} CH_3CH_2CH{=}\overset{+}{O}{-}H \xrightarrow{Cl^-} CH_3CH_2CH{=}O + HCl$

1f. $H_2C{=}C(CH_3){-}CH_2CH_2CH_2{-}C(CH_3){=}CH_2 + H{-}OSO_3H \longrightarrow$ tertiary carbocation $\longrightarrow$ cyclized carbocation $\xrightarrow{^-OSO_3H}$ H_2SO_4 + 1,5,5-trimethylcyclohexene (ring with H_3C, CH_3, CH_3)

or

$^{\ominus}OSO_3H$

H H

H_3C ⊕ CH_3 CH_3 ⟶ H_3C CH_3 CH_3 + H_2SO_4

1g. H–C≡C–OCH_3 ⟶ H_2C=C=$\overset{\oplus}{O}CH_3$ ⟶ H_2C=C(OCH_3)–$\overset{\oplus}{O}H_2$

Cl–H $Cl^{\ominus}$ OH_2 $Cl^{\ominus}$

HCl + $H_3C\overset{O}{\overset{\|}{C}}OCH_3$ ⟵ $H_3C\overset{\overset{\oplus}{O}H}{\overset{\|}{C}}OCH_3$ $Cl^{\ominus}$ ⟵ H_2C=C(OCH_3)OH

Cl–H

1h. (2-ethoxycyclohexanone, H, $^{\ominus}OH$, OC_2H_5) ⟶ (enolate, OC_2H_5) + H_2O ⟶ (cyclohexenone) + $^{\ominus}OC_2H_5$

This reaction cannot proceed by a one-step E2 elimination, because the ethoxide anion is not a good leaving group. It requires formation of a negative charge on the β carbon atom to eject the ethoxide anion.

1i. (4-chlorocyclohexanone, H, Cl, $^{\ominus}OH$) ⟶ (H, H, $^{\ominus}OH$) + H_2O + $Cl^{\ominus}$ ⟶ (carbanion ⊖) H–OH ⟶ (cyclohexenone) + $^{\ominus}OH$

1j. As in problem *h*, the ethoxide anion is too poor a leaving group to be eliminated in an E2 process. Furthermore, formation of the enolate anion cannot result in elimination of a leaving group at a carbon γ to the carbonyl group. Thus, no elimination reaction will take place.

1k. $CH_3-C(OH)(CH_3)-CH_2CCH_3$ (with $\overset{\ominus}{O}H$; C=O on the ketone) $\longrightarrow$ $CH_3-C(O^{\ominus})(CH_3)-CH_2CCH_3$ + H_2O $\longrightarrow$ CH_3CCH_3 (O) + $\overset{\ominus}{C}H_2-CCH_3$ (O) (HO–H)

$\downarrow$

$2\ CH_3CCH_3$ (O)

1l. $C_6H_5-C(OH)(CH_3)-CH_2CH$ (O) (with $\overset{\ominus}{O}H$) $\longrightarrow$ H_2O + $C_6H_5-C(O^{\ominus})(CH_3)-CH_2CH$ (O)

$\longrightarrow$ $C_6H_5-CCH_2-H$ (O) + $\overset{\ominus}{C}H_2-CH$ (O)

$\longrightarrow$ $C_6H_5-C\overset{\ominus}{C}H_2$ (O) + CH_3CH (O)

$\longrightarrow$ $C_6H_5-CCH_2CHCH_3$ (O, $O^{\ominus}$) $\xrightarrow{H_2O}$ $C_6H_5-CCH_2CHCH_3$ (O, OH)

1m.

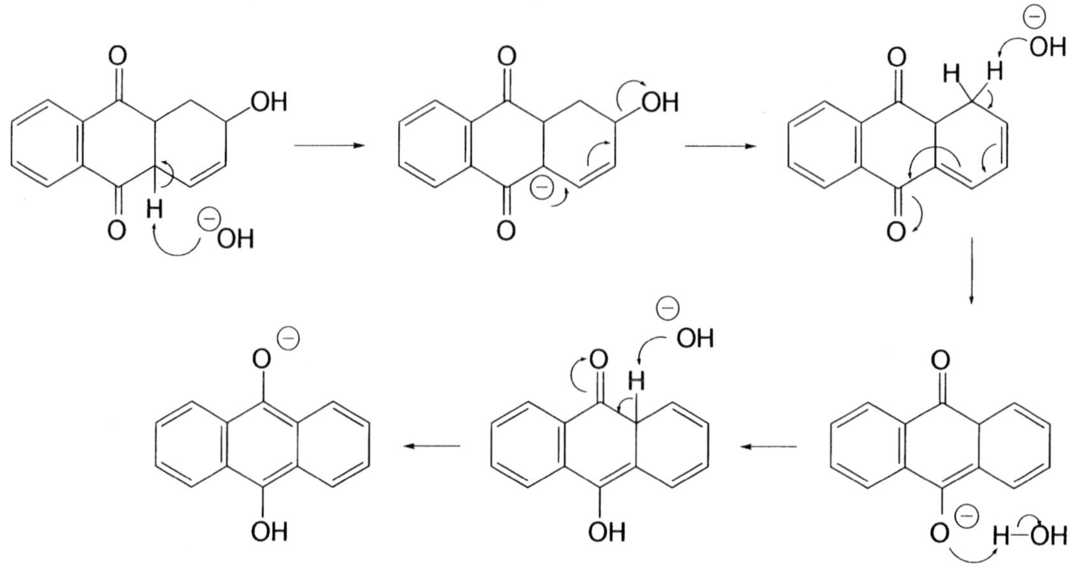

1n. $CH_3-C(OH)(C{\equiv}N)-CH{=}CH_2 \rightarrow CH_3-C(O^-)(C{\equiv}N)-CH{=}CH_2 \rightarrow CH_3C(=O)-CH{=}CH_2 + {}^-CN \rightarrow CH_3C(=O)-\bar{C}H-CH_2C{\equiv}N \rightarrow CH_3O^- + CH_3C(=O)CH_2CH_2C{\equiv}N$

HOCH3

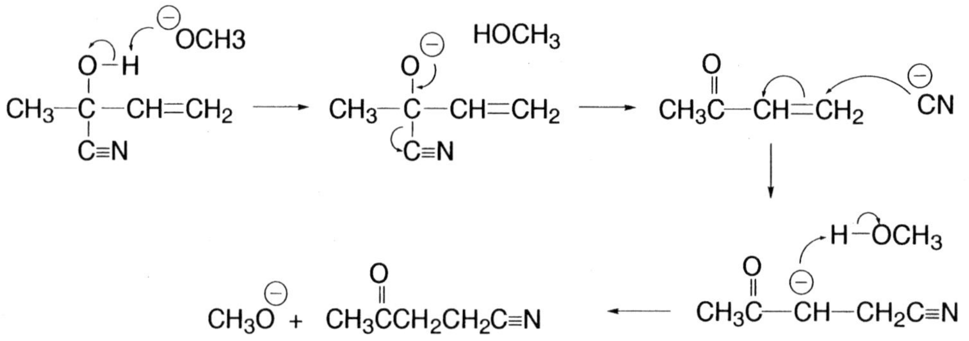

1o.

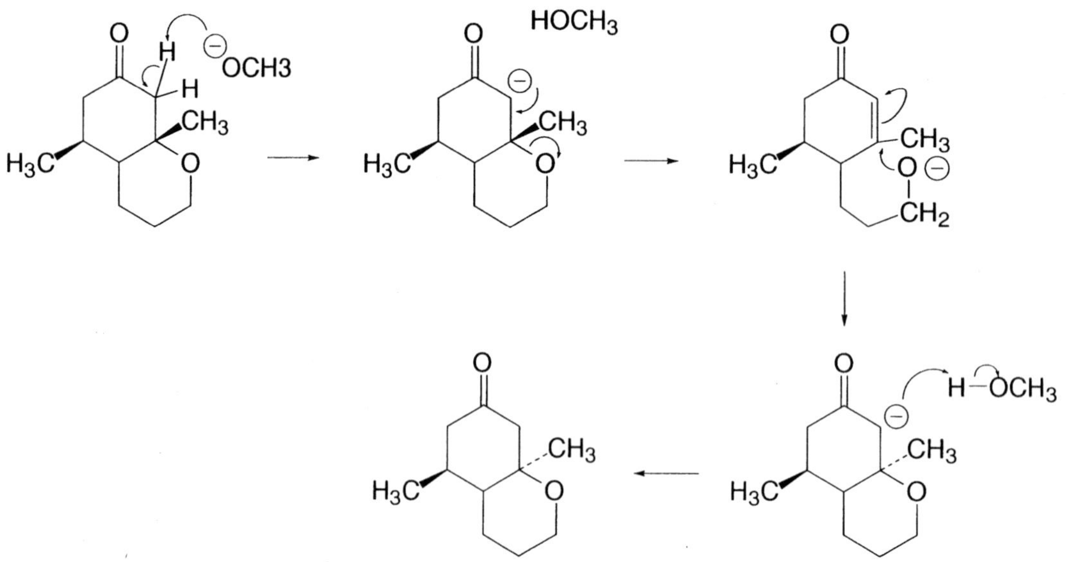

2a. Compound **A** is a relatively strong acid, compared to other phenols, because the negative charge in its anion can be distributed onto oxygens of the nitro group.

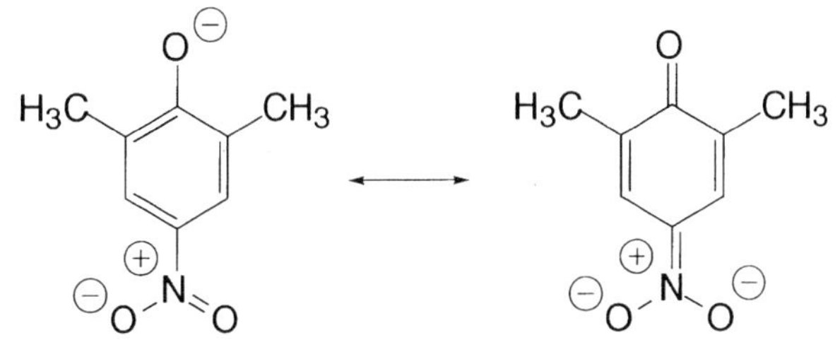

Anion of **A**

In order for that to happen, however, the nitro group must be in, or nearly in, the same plane as the aromatic ring. In phenol **B** steric interferences between the methyl groups and the oxygens of the nitro group would prevent the nitro group from being in the same plane as the aromatic ring (or would make the planar structure one of relatively high energy.) Thus, stabilization of the anion of **B** is reduced by steric inhibition of resonance.

2b. Two factors combine to make amine **E** a much weaker base than **C**. (1) The nitrogen atom is bonded to an sp^2 hybridized carbon in **E**, but only to sp^3 carbons in **C**. The low energy sp^2 orbital is more strongly electron-attracting than sp^3 orbitals, and its inductive effect would make it harder to protonate **E** than **C**. (2) There is a significant resonance interaction between the nitrogen atom and the aromatic ring in **E**. This resonance stabilization is lost if **E** is protonated, so that the protonated form of **E** is higher in energy, relative to the neutral amine, than is the case with amine **C**.

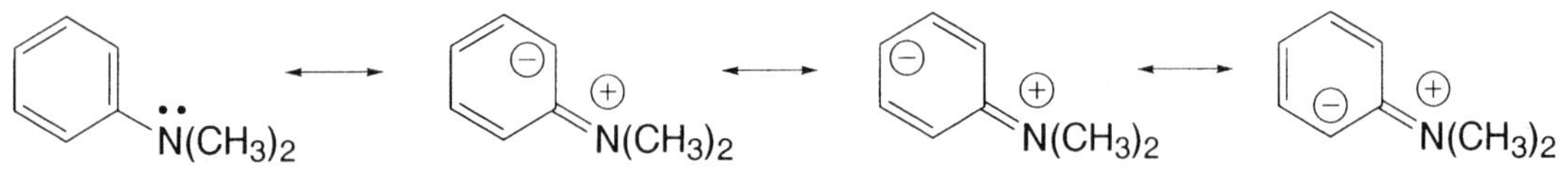

E

The inductive effect of the aromatic ring decreases the basicity of **D** compared to **C**, just as it does that of **E**. However, the rigid bicyclic ring structure in **D** holds the unshared electron pair on nitrogen in a position essentially orthogonal to the π-orbitals of the aromatic ring. This should effectively eliminate any resonance stabilization of **D**. Thus, **D** is a weaker base than **C**, but a stronger base than **E**.

CHAPTER 2

ELECTROCYCLIC REACTIONS

PROBLEM SOLUTIONS

1a. The starting triene first undergoes rotation around single bonds to convert it to a conformation which can undergo an electrocyclic reaction. The product shown results from a conrotatory cyclization involving six π electrons. The cyclization is forbidden as a thermal process but is allowed as a photochemical process.

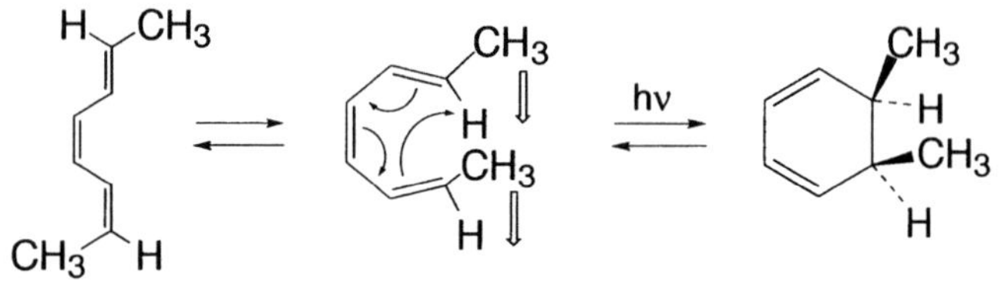

b. This is a disrotatory reaction. Since four electrons are involved, the reaction is allowed if the starting material is subjected to photoirradiation, but forbidden under purely thermal conditions.

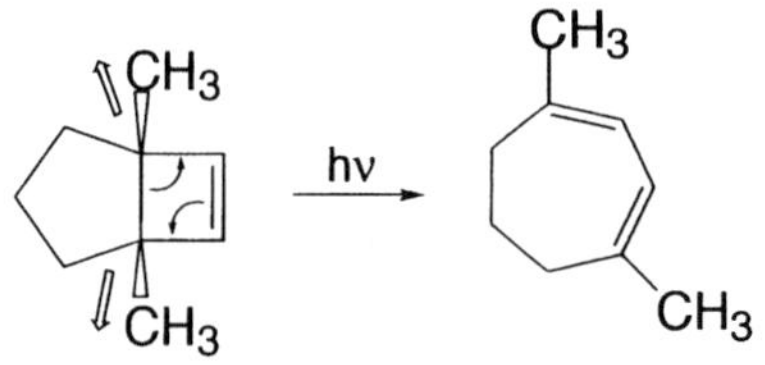

2. Since the cyclobutene ring is highly strained, the diene form is thermodynamically more stable than the cyclized form. Thus, in the absence of photoirradiation, the equilibrium mixture should consist primarily of the diene.

The situation is more complicated when the mixture is irradiated, and depends on the wave length of the light. 'Near UV' light, including wave lengths around 225 nm, will be absorbed by the diene and thus drive the equilibrium toward the cyclobutene side. On the other hand, 'far UV' light, including wave lengths around 180 nm but not around 225 nm, will be absorbed by the cyclized form, and drive the reaction even further toward the diene form than in the absence of irradiation.

Photochemical reactions resulting from irradiation by far UV light are much less common than those resulting from irradiation by near UV light, since glass and most solvents absorb light in the far UV region. Thus, unless irradiation by light in the far UV region is specified, problems involving photoirradiation should be assumed to imply irradiation by light with wave lengths above 210 nm.

3.

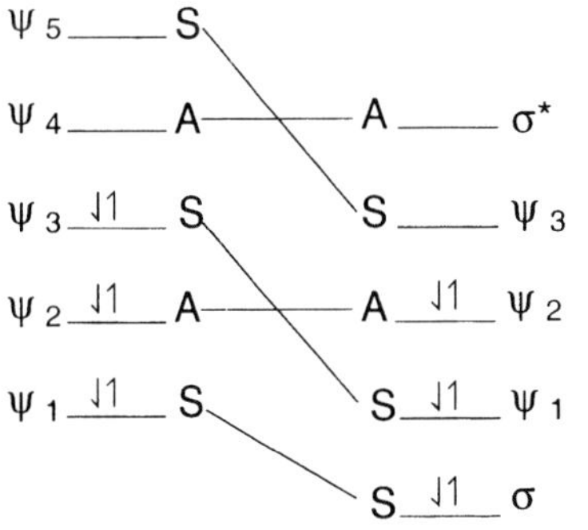

Correlation diagram for the disrotatory cyclization of a pentadienyl anion, showing symmetries around the mirror plane. This is an allowed reaction.

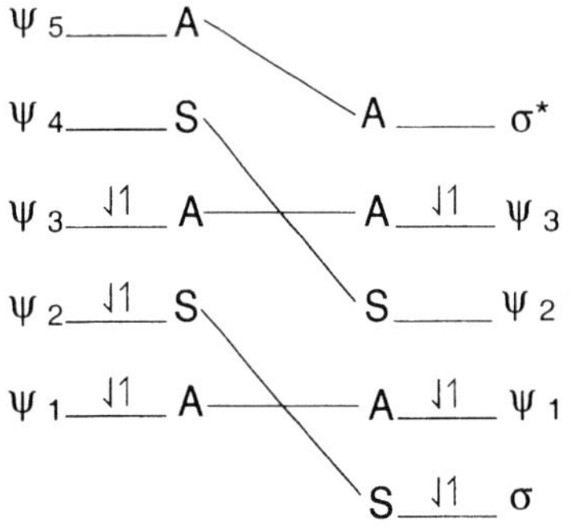

Correlation diagram for the conrotatory cyclization of a pentadienyl anion, showing symmetries around the twofold axis of rotation. This is a forbidden reaction.

Thermal cyclization of the anion shown in Eq. (3a) should therefore proceed in a disrotatory manner.

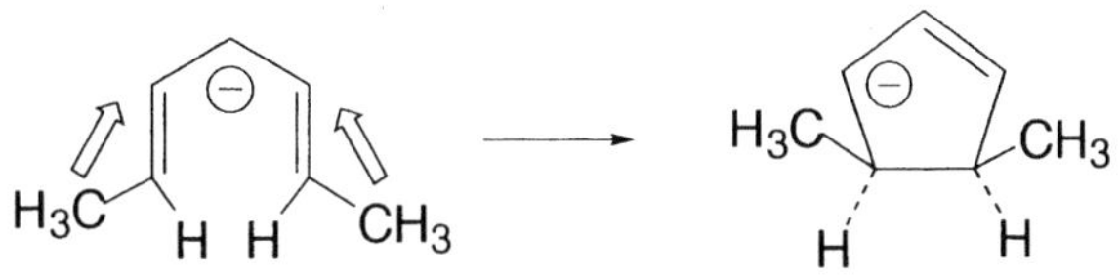

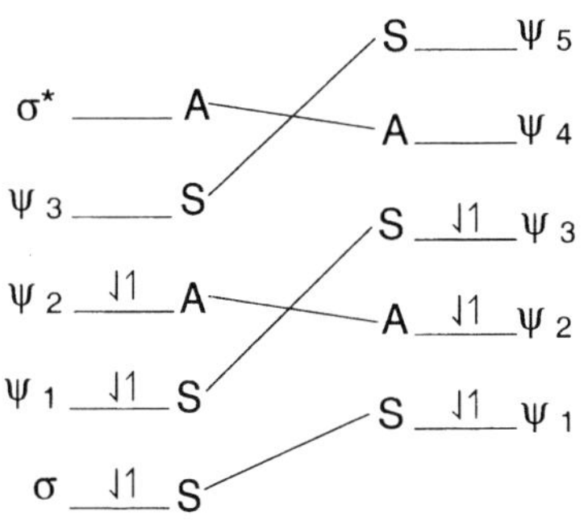

Correlation diagram for the disrotatory ring opening of a cyclopentenyl anion.

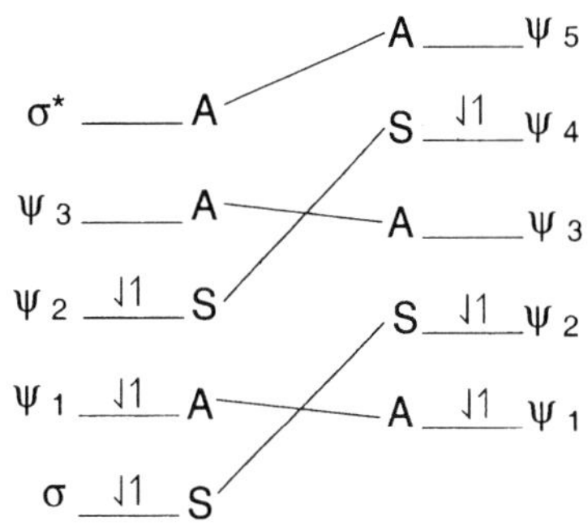

Correlation diagram for the conrotatory ring opening of a cyclopentenyl anion.

The thermal ring opening of the anion shown in Eq. (3b) should therefore proceed in a disrotatory manner.

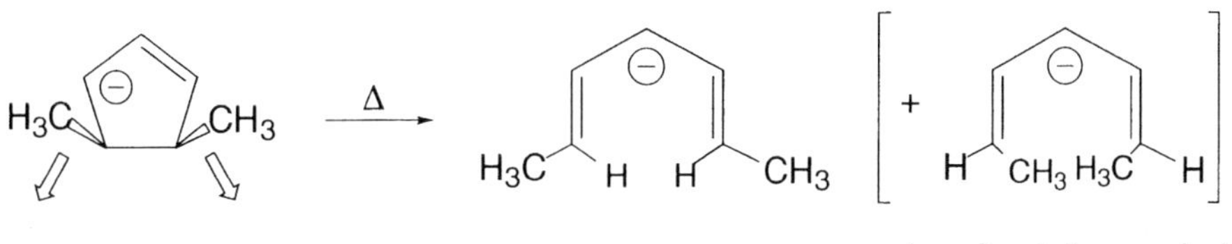

minor disrotation product

4a.

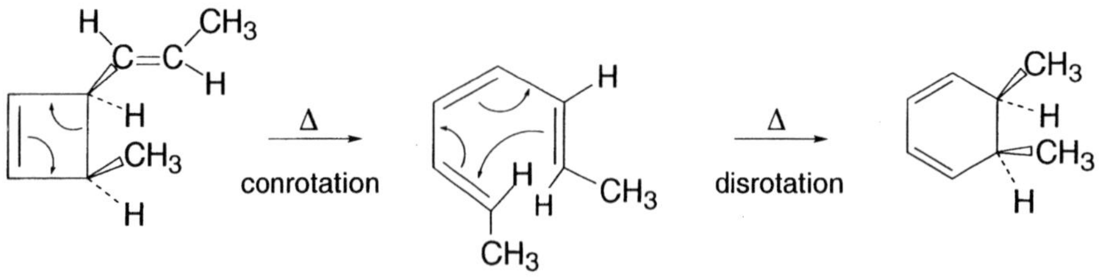

H C=C CH3 H H CH3 H

Δ conrotation

H H H CH3 CH3

Δ disrotation

CH3 H CH3 H

Note that only one of the two possible conrotation products from the first step has the proper geometry to form a six-membered ring in the second step.

4b.

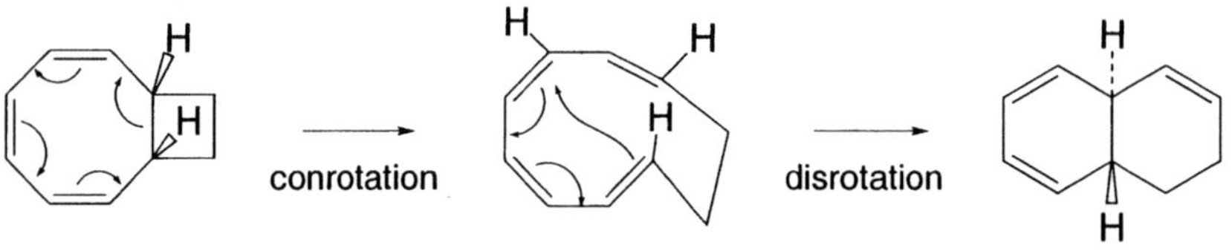

H H

conrotation

H H H

disrotation

H H

4c.

CH3 ⊖ OR H D C H C Br CH3

"anti" elimination

CH3 H D H CH3

CH3 H3C H D

CH3 H H CH3 D

CH3 ⊖ OR H D C CH3 C Br H

CH3 H H CH3 H

CH3 H CH3 H

Δ

CH3 H CH3 H

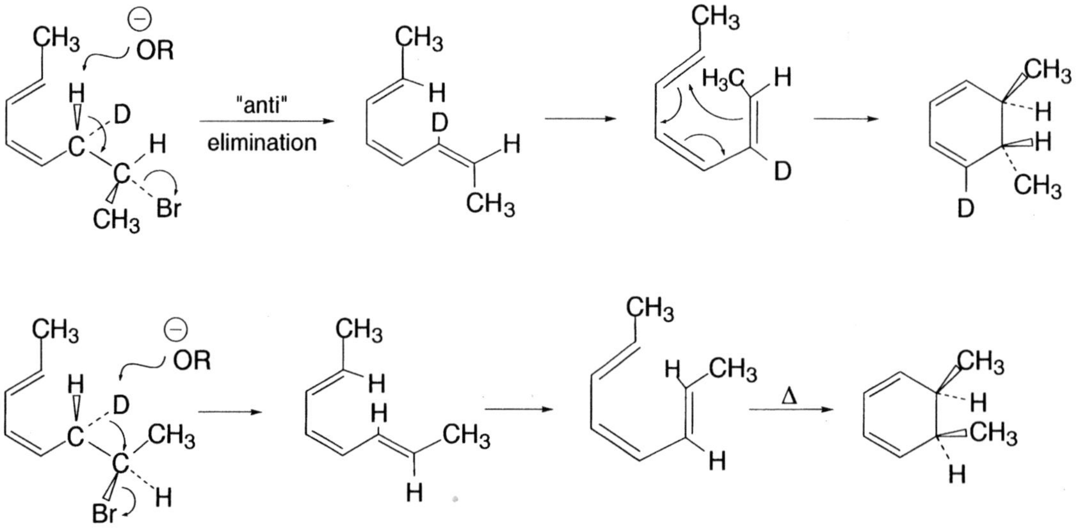

4d.

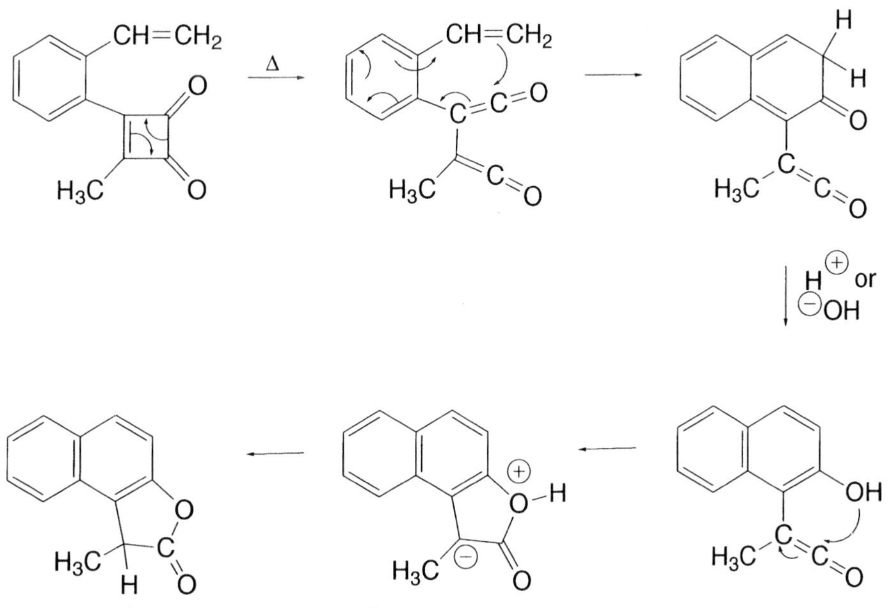

Δ

$H^{\oplus}$ or $^{\ominus}OH$

4e.

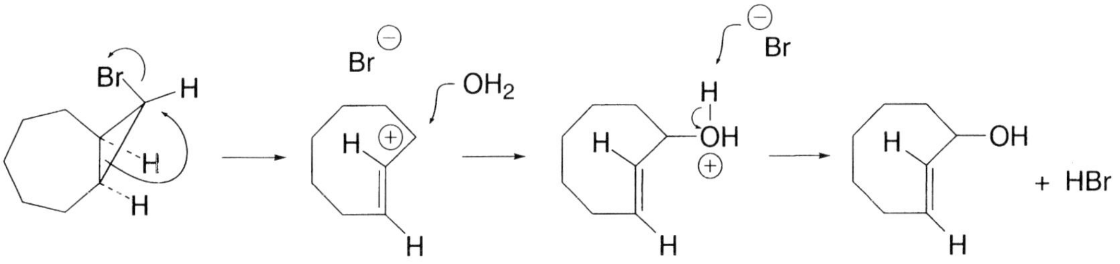

+ HBr

4f.

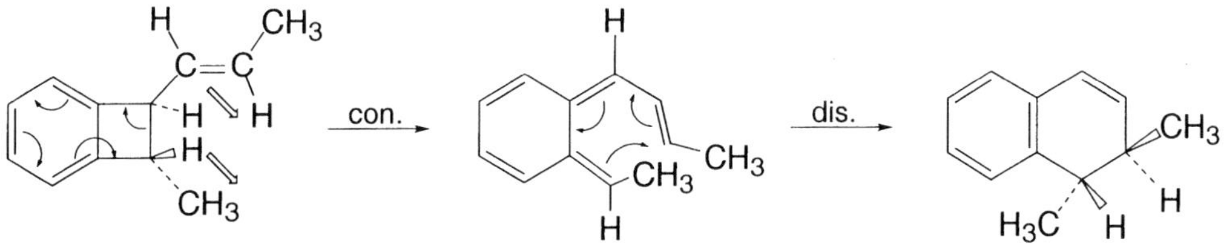

con.

dis.

5a.

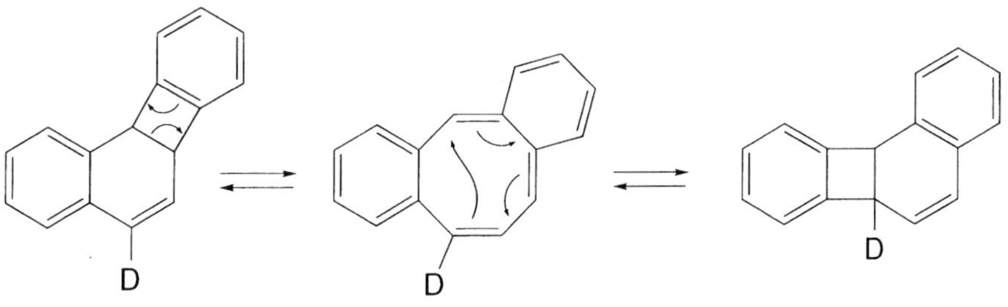

5b.

$$\xrightarrow{\Delta}$$

5c.

$CH_2=C(C_4H_9)-C\equiv C-CH_3 \xrightarrow{h\nu}$ [cyclobutadiene, CH_3, C_4H_9] $\longrightarrow C_4H_9-C\equiv C-C(=CH_2)CH_3$

CHAPTER 3

CYCLOADDITION AND CYCLOREVERSION REACTIONS

PROBLEM SOLUTIONS

1a. This reaction is an eight electron cycloaddition. It would be forbidden as a concerted, suprafacial process, but an antarafacial cycloaddition would be allowed.

b. This reaction is a six electron cycloaddition. It would be allowed as a concerted, suprafacial process. (An antarafacial reaction would be extremely unlikely, because of the very strained geometry of the transition state.)

c. This reaction is a six electron, suprafacial cycloreversion. (Note that one π bond does not take part in the reaction.) It is allowed by orbital symmetry conservation rules.

d. This reaction is a six electron, suprafacial cycloaddition. It is allowed.

e. This reaction is an eight electron cycloaddition. As a photochemical reaction, the suprafacial addition is allowed, but an antarafacial addition would be forbidden.

f. This reaction can most simply be regarded as a four electron, suprfacial photochemical cycloreversion, and is allowed. (Note that if the ring is split by cleaving the bonds linking the pairs of methyl *cis* to each other, the reaction would actually be a doubly antarafacial, $[\sigma 2_a + \sigma 2_a]$, process. However, that would also be allowed as a photochemical reaction.)

2a.

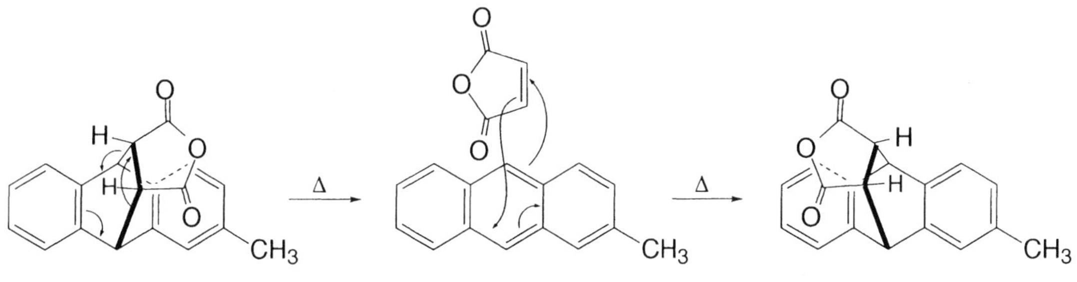

2b.

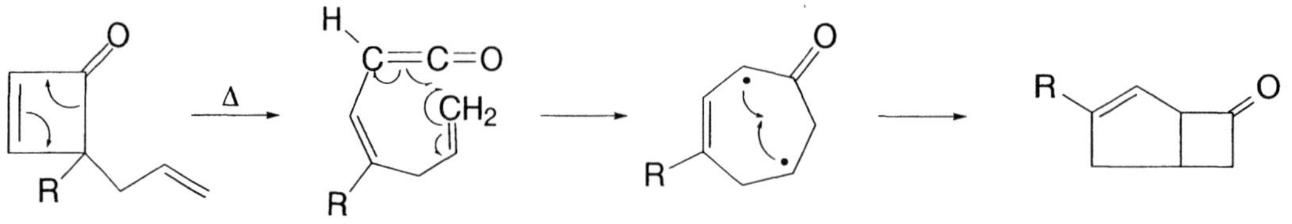

H
C=C=O
CH2
O
R
Δ

(Note, however, the uncertainty in regard to the mechanisms of cycloadditions of ketenes discussed in the text.)

2c.

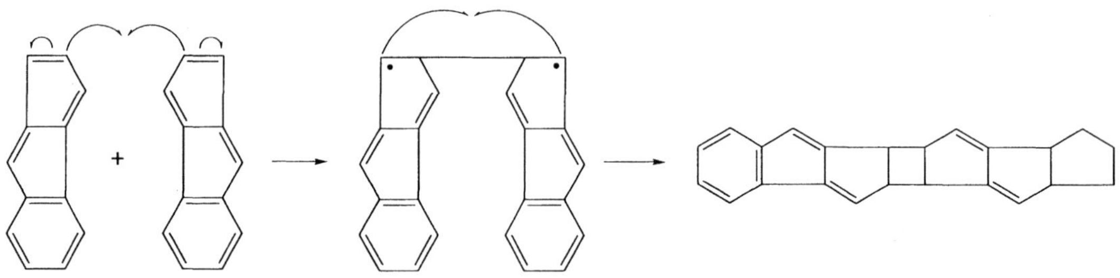

+

(A concerted cycloaddition is a forbidden reaction.)

2d. The two five-membered rings in the starting material in problem (c) comprise a closed, cyclic system containing eight π electrons. By itself, this eight carbon structure, which has been given the trivial name *pentalene*, should be a high energy, antiaromatic system. The starting material in problem (c), benz[b]pentalene, contains an aromatic ring, but overall comprises an antiaromatic, twelve π electron system. (This can be demonstrated by drawing the other Kekulé resonance form for the aromatic ring.) Combination of two benz[b]pentalene molecules to form a diradical leaves the aromatic rings of both molecules intact, but eliminates their antiaromatic character. That presumably accounts for this "thermal" dimerization reaction taking place at such a low temperature.

2e.

Δ

$H-O-SO_3H$

$^{\ominus}OSO_3H$

H_2SO_4

$-H^{\oplus}$

OH_2

OH

2f.

Δ

CO_2CH_3

CH

CH_2

COCH$_3$

2g.

CH_2 CH_2

CH_2-CH_2

$CH_2\cdot$ $\cdot CH_2$

CH_2-CH_2

2h. Aco H₃C H CH₃ R O CH HC C Δ Aco H₃C H O CH CH₃ R Δ $[_{\sigma}2_s + _{\pi}2_s + _{\sigma}2_s]$ CH₃ HO O CH CH₃ R

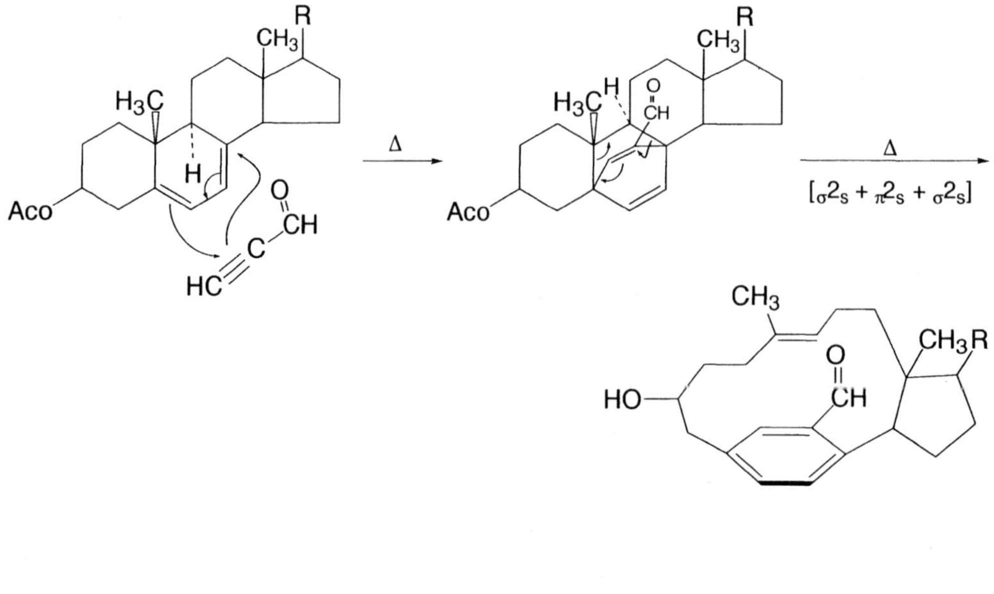

2i. O hν O CH CH₂ O O O C₂H₅ O •CH₂ H

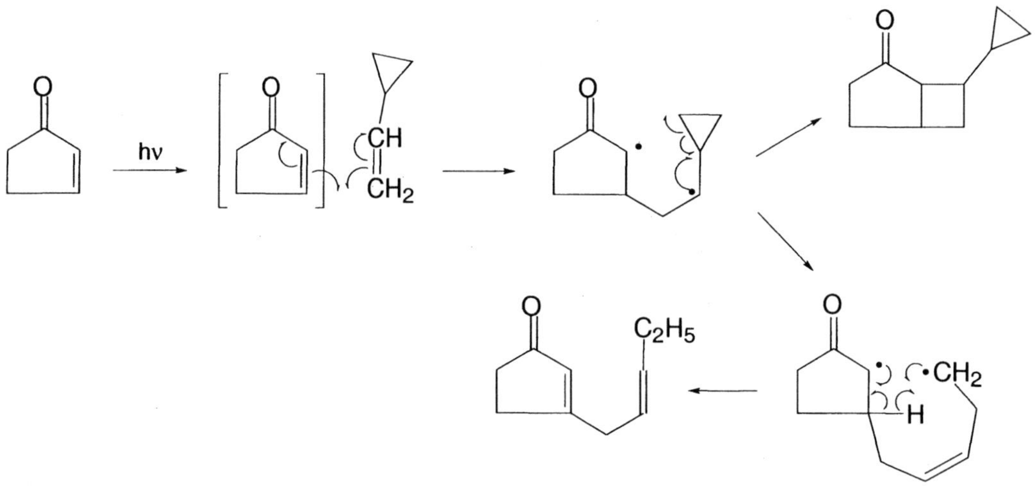

2j. For simplicity, the "end" aromatic rings of each anthracene or dihydroanthracene unit have been omitted in the equations below. Only the central, reacting rings are shown.

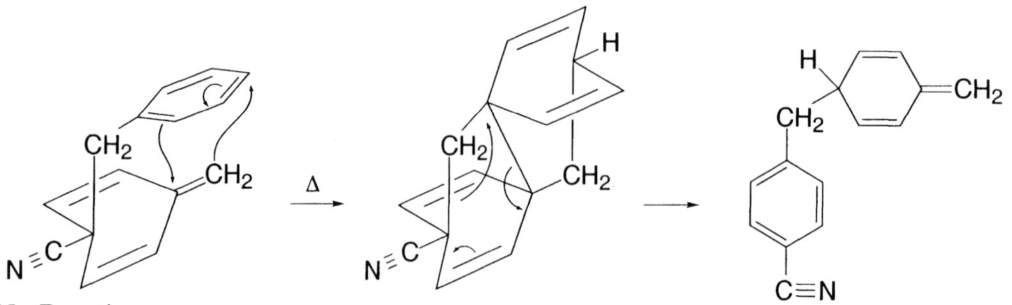

CH₂ CH₂ N≡C Δ H CH₂ CH₂ N≡C H CH₂ CH₂ C≡N

3a. No Reaction.

3b. 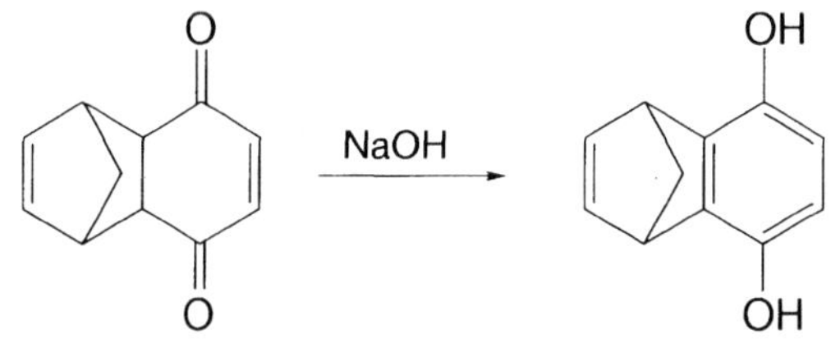

A **B**

3c. 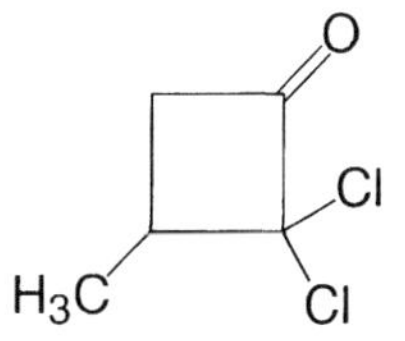

3d. 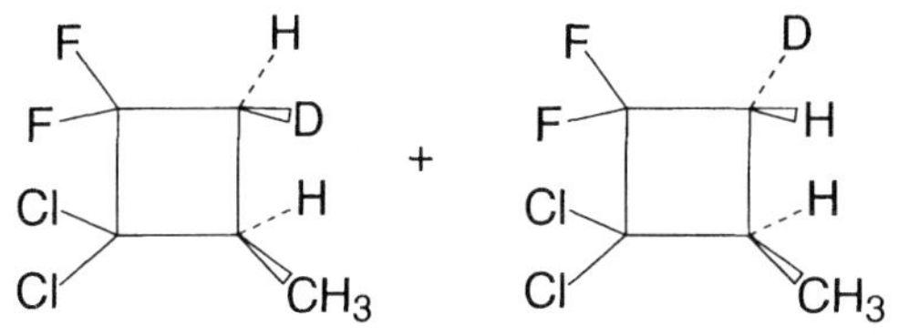

3e. 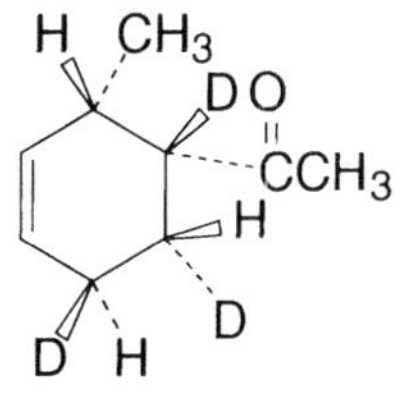

3f. 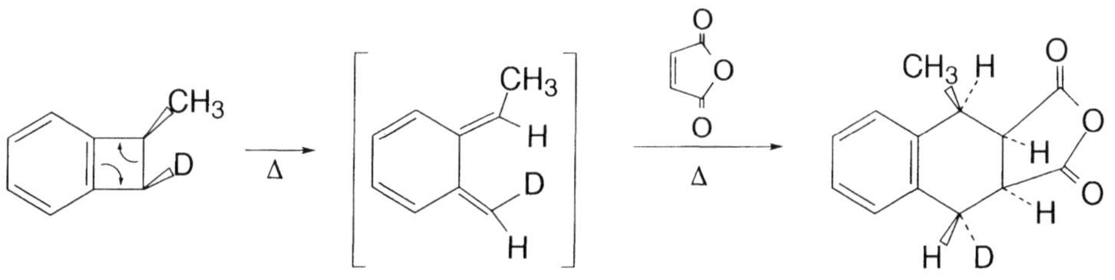

CHAPTER 4

SIGMATROPIC REACTIONS

PROBLEM SOLUTIONS

1. To employ the frontier orbital approach, imagine the transition state for the [3,5] shift to resemble an allyl radical and a pentadienyl radical with their ends in close proximity to each other. The molecular orbitals for the two radicals are shown below.

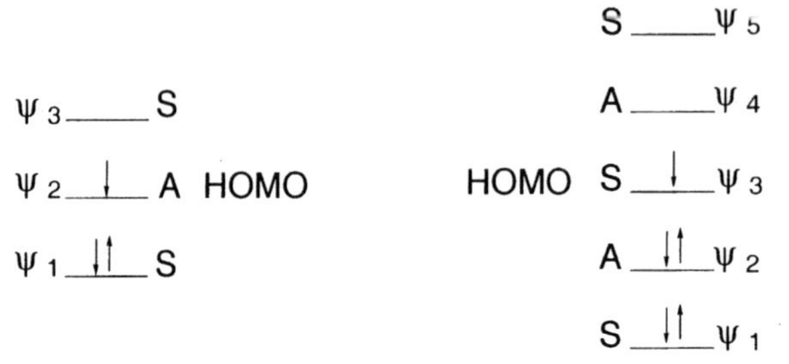

MO's of an allyl radical		MO's of a pentadienyl radical
		S ___ ψ_5
ψ_3 ___ S		A ___ ψ_4
ψ_2 _↓_ A HOMO		HOMO S _↓_ ψ_3
ψ_1 _↓↑_ S		A _↓↑_ ψ_2
		S _↓↑_ ψ_1

The HOMO, ψ_2, of the allyl radical is antisymmetric and the HOMO, ψ_3, of the pentadienyl radical is symmetric. Therefore the thermal rearrangement is forbidden as a suprafacial process but, in principle, would be allowed as an antarafacial process. In a photochemical reaction, the electron in one of the HOMO's would be raised to the next higher orbital, which would thus become the new HOMO. The HOMO's of the two radical units would thus both by symmetric or would both be antisymmetric. The suprafacial rearrangement would be allowed, and the antarafacial rearrangement would be forbidden.

2a. This reaction is a [1,3] shift. As a suprafacial, photochemical, 4 electron process, it is theoretically allowed.

2b. This reaction is a [1,5] shift. As a suprafacial, thermal, 6 electron process, it is theroetically allowed.

2c. This reaction is a [3,3] shift. As a suprafacial, thermal, 6 electron process, it is theoretically allowed.

2d. This reaction is a [3,3] shift. As a suprafacial, photochemical, 6 electron process, it is theoretically forbidden.

2e. This reaction involves a [3,5] shift, which is theoretically forbidden as a thermal, suprafacial, 8 electron process.

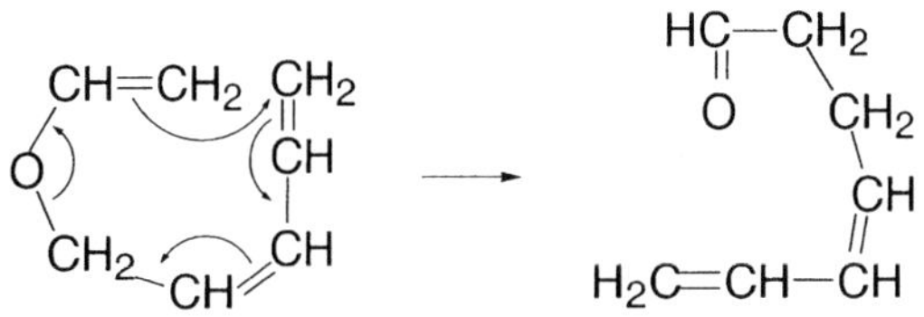

CH=CH₂ CH₂ / O / CH / CH₂ CH CH ⟶ HC—CH₂ / O CH₂ / CH / H₂C=CH—CH

3. The products of these reactions would be:

3a. 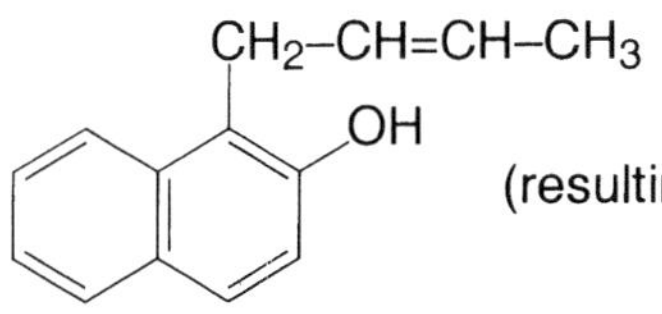 CH_2–CH=CH–CH_3, OH (resulting from an *o*-Claisen rearrangement)

3b. 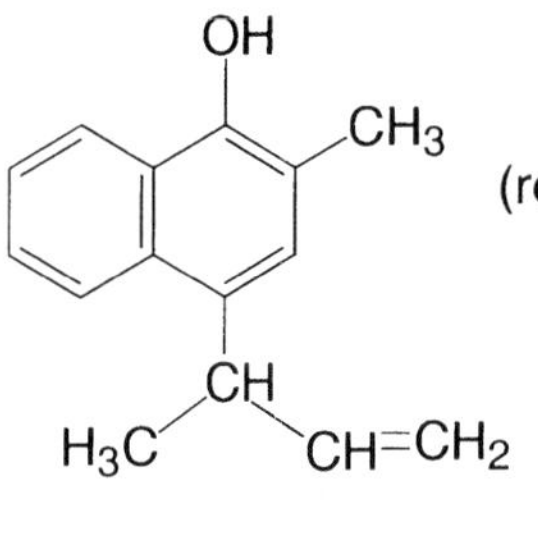 OH, CH_3, CH, H_3C, CH=CH_2 (resulting from a *p*-Claisen rearrangement)

3c. An initial [1,5] phenyl shift should yield diene A. However, [1,5] hydrogen migrations would partially convert the initial product to the other isomers shown below.

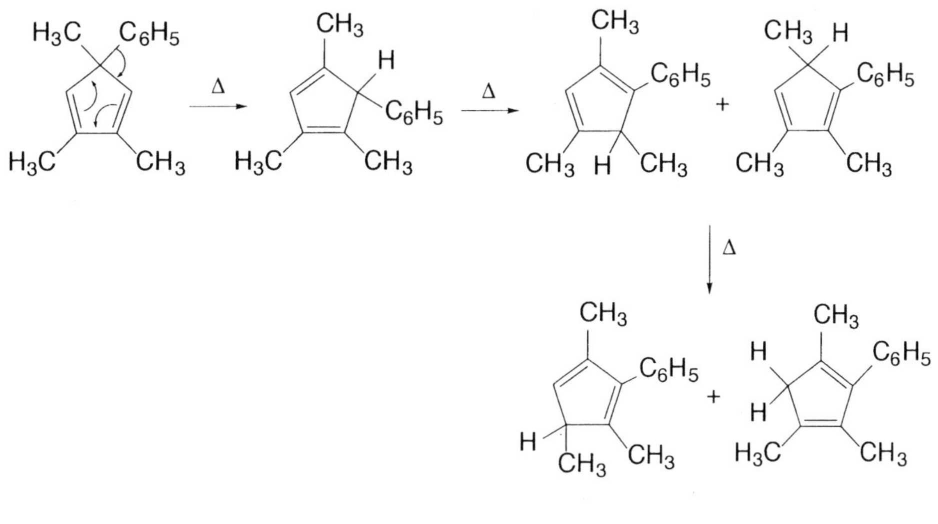

H_3C C_6H_5 / H_3C CH_3 —Δ→ CH_3 H C_6H_5 / H_3C CH_3 —Δ→ CH_3 C_6H_5 / CH_3 H CH_3 + CH_3 H C_6H_5 / CH_3 CH_3

↓ Δ

CH_3 C_6H_5 / H CH_3 CH_3 + CH_3 H C_6H_5 / H H_3C CH_3

3d. 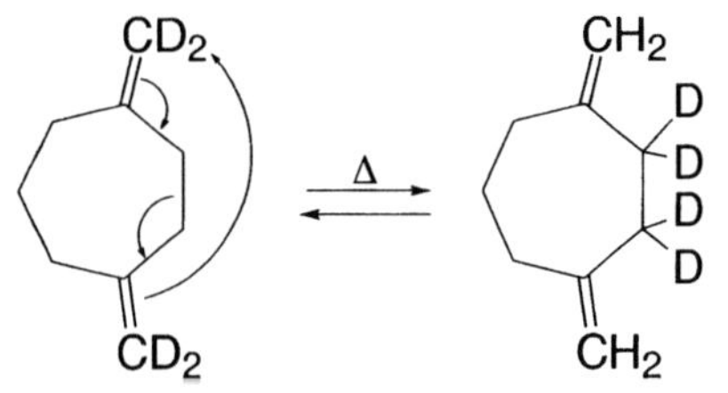 (a Cope rearrangement)

3e. 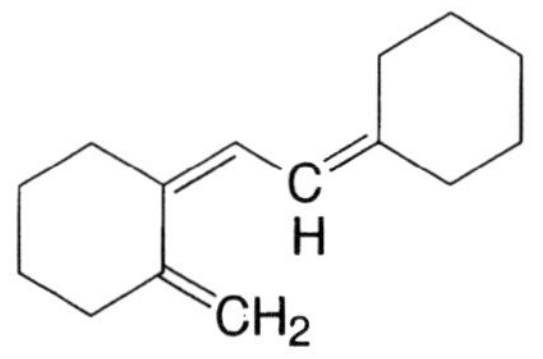 (resulting from a [1,5] hydrogen shift)

3f.

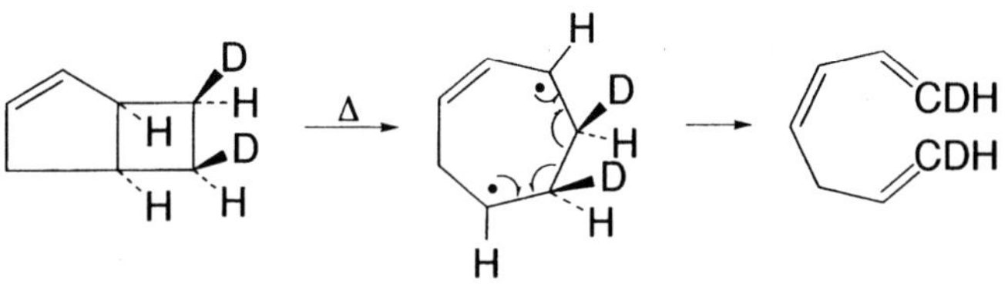

Since a concerted [$\sigma 2_S + \sigma 2_S$] ring opening is forbidden, the only reasonable reaction would be ring opening by a diradical process, via the most stable diradical. A mixture of the four possible stereoisomers of the resulting triene should be formed.

4a.

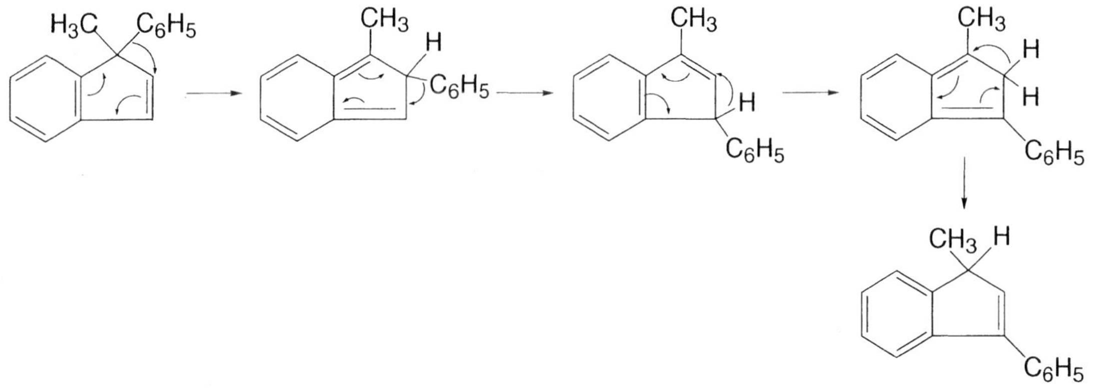

H3C C6H5 → CH3 H C6H5 → CH3 H C6H5 → CH3 H H C6H5 → CH3 H C6H5

4b.

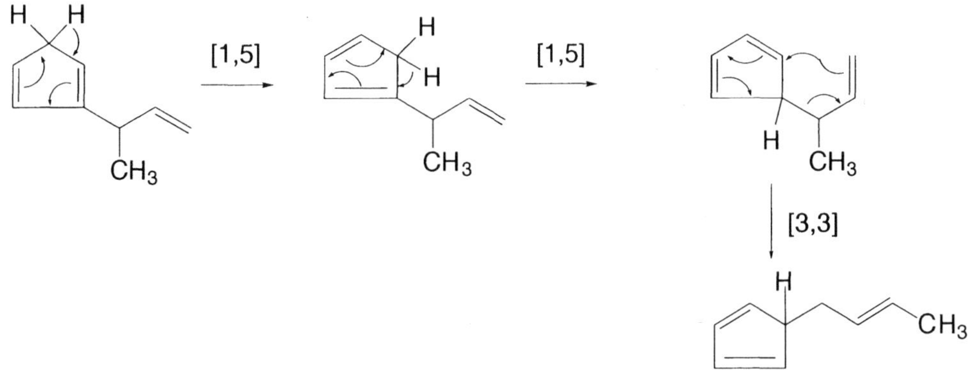

H H → [1,5] → H H → [1,5] → H CH3 → [3,3] → H CH3

CH3 CH3

4c.

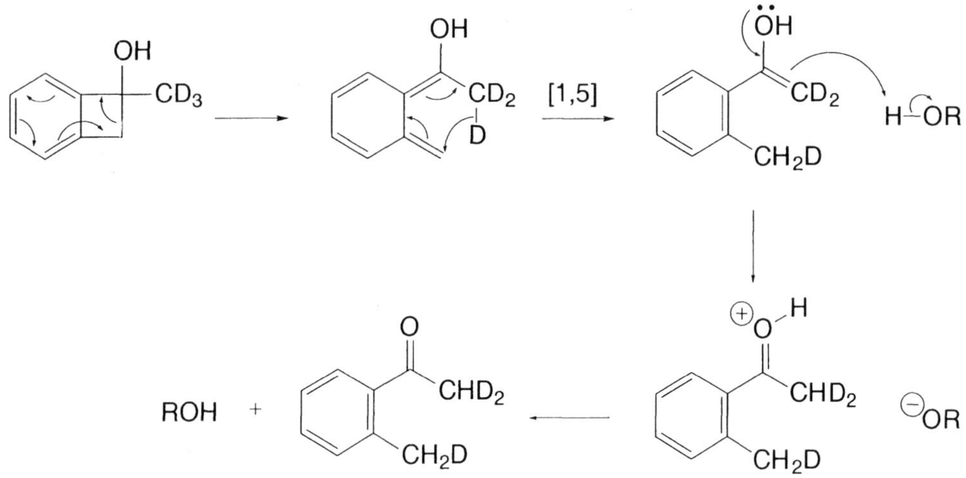

OH CD3 → OH CD2 D → [1,5] → ÖH CD2 CH2D H–OR → ⊕O H CHD2 CH2D ⊖OR → ROH + O CHD2 CH2D

4d.

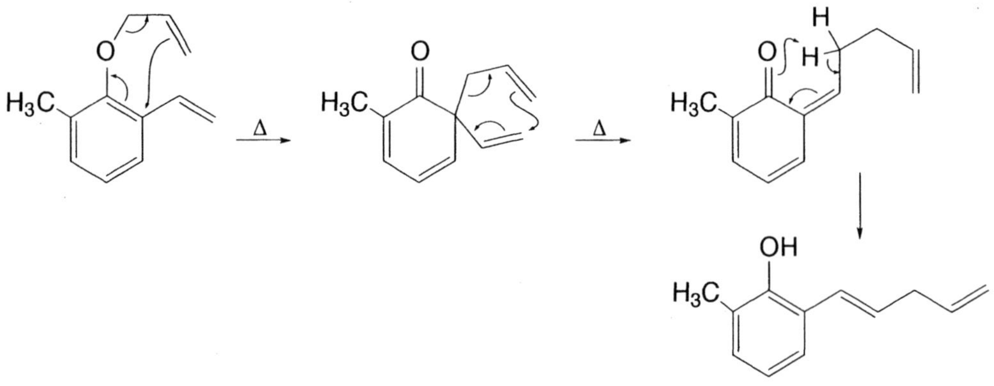

The third step might also proceed via a sequence of nonconcerted, acid or base-catalyzed reactions.

4e.

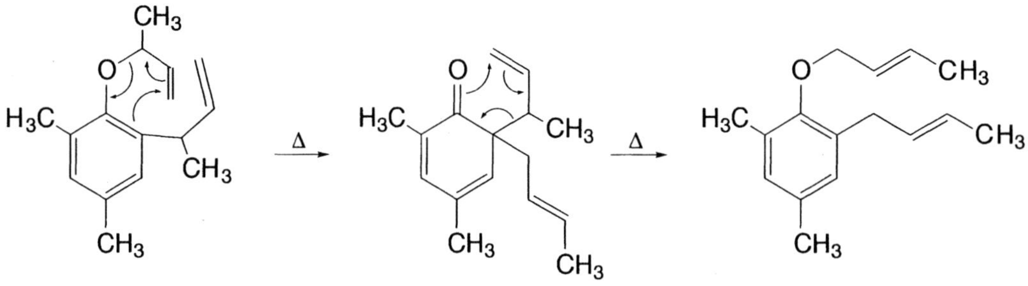

4f.

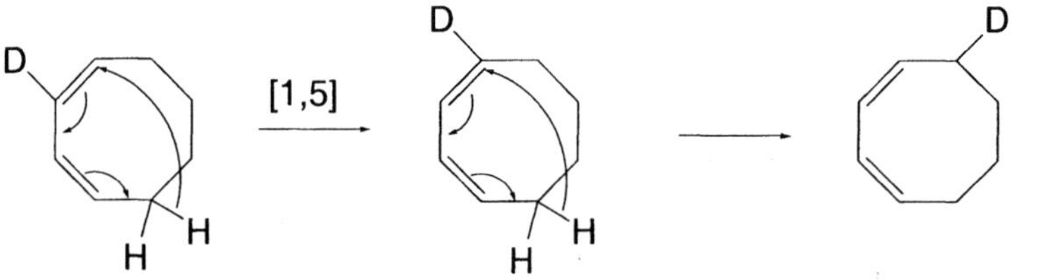

4g.

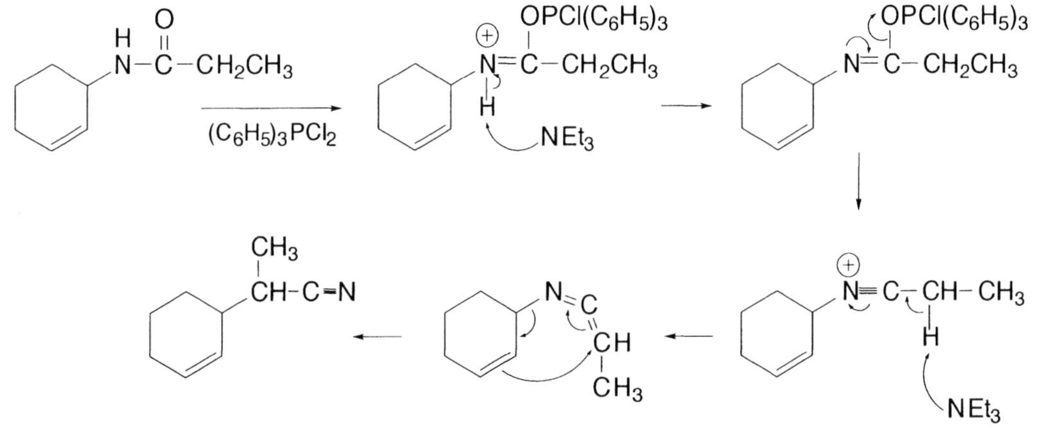

The timing of the first few steps in this reaction is not certain, and several reasonable mechanistic sequences can be suggested. The important point is to arrive at the ketenimine, so that a [3,3] migration in the final step will form the observed product.

4h.

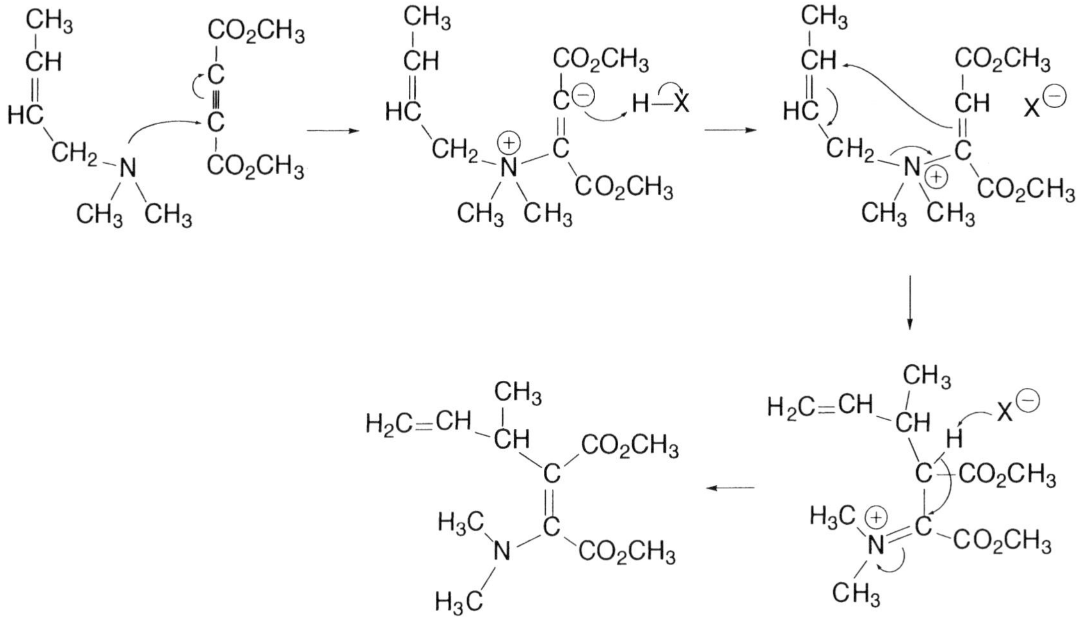

4i.

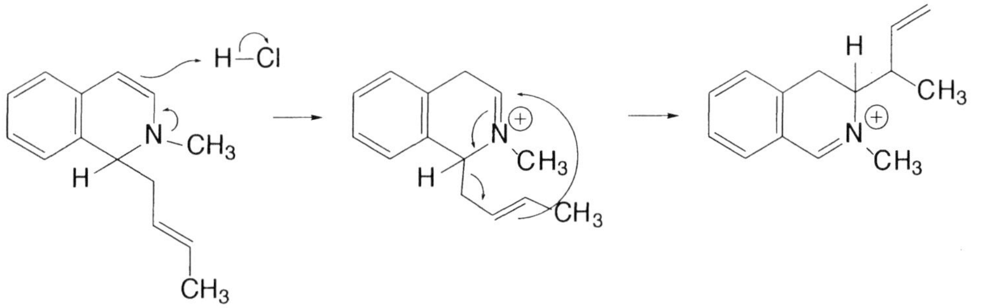

4j. CH_2CH_3 CH_2CH_3 → CH_2CH_3 $CH-CH_3$ H $\xrightarrow[\text{H shift}]{[1,7]}$ CH_2CH_3 CH_3 CH_3 → CH_2CH_3 CH_3 CH_3

CHAPTER 5

LINEAR FREE ENERGY RELATIONSHIPS

PROBLEM SOLUTIONS

1. The relative rates of saponification of ethyl benzoate and ethyl 3,4-dichlorobenzoate can be estimated by use of the Hammett equation.

   This reaction does not result in a charge being directly placed on the aromatic ring. Therefore, the effects of substituents can be calculated by use of their standard σ values. Since there are substituents in both the *meta* and *para* positions, the overall substituent constants would be obtained by adding σ_m and σ_p. According to Table 2, the value of ρ for the saponification of ethyl benzoates under the conditions specified in this problem is 2.547.

   Therefore, $\log k/k_H = 2.547\ (0.37 + 0.23) = 1.5282$

   where k_H is the rate of saponification of ethyl benzoate and k is the rate of saponification of ethyl 3,4-dichlorobenzoate

   $k/k_H = \text{antilog}\ 1.5282 = \text{ca. } 33.7$

2a. The rate limiting step in this reaction is the addition of a cyanide anion to the carbonyl group.

$$\mathrm{ArCH{=}O} + {}^{\ominus}\mathrm{CN} \longrightarrow \mathrm{ArCH(O^{\ominus})\text{-}C{\equiv}N}$$

Since this step should be accelerated by electron-withdrawing groups on the aromatic ring, its Hammett plot should have a positive slope. Since the negative charge is not directly distributed onto the ring, the best plot should be obtained using values of σ.

b. Electron-withdrawing groups on the aromatic ring should raise the energy of the starting material, and should lower the energy of the product and of the transition state leading to the product.

$$ArN^{\oplus}(CH_3)_3 + {}^{\ominus}OCH_3 \longrightarrow ArN(CH_3)_2 + CH_3OH$$

Therefore, electron-withdrawing groups should accelerate the reaction. Its Hammett plot should have a positive slope. Since the nitrogen atom in the product, and, presumably, in the transition state, can donate electrons directly to the ring, it seems probable that the best Hammett plot would be obtained using values of σ^-.

c. This is an S_N1 reaction, in which the rate limiting step is formation of a carbocation. That step would be inhibited by electron-withdrawing substituents on the rings.

$$Ar_3C-\overset{\oplus}{O}H_2 \xrightarrow{slow} H_2O + Ar_3C^{\oplus} \xrightarrow[fast]{Cl^{\ominus}} Ar_3C-Cl$$

Therefore, a Hammett plot should have a negative slope. Since the positive charge can be distributed onto the aromatic rings, the best plot should be obtained using values of σ^+.

3. This reaction may reasonably proceed by a two step mechanism:

$$1)\ ArCOOH + \phi_2C{=}\overset{\oplus}{N}{=}\overset{\ominus}{N} \longrightarrow ArCOO^{\ominus} + \phi_2CH-\overset{\oplus}{N}{\equiv}N$$

$$2)\ ArCOO^{\ominus} + Ar_2CH-\overset{\oplus}{N}{\equiv}N \longrightarrow ArCOO-CH\phi_2 + N_2$$

The value of ρ for the first step should be positive, and for the second step negative. The overall value of ρ would be dependent on the ρ values for both steps. The fact that the overall value of ρ is positive indicates that the absolute value of ρ for the first step is larger than for the second step.

Formation of the carboxylate anion in methanol is assisted by the polarity of the solvent and by hydrogen bonding between the solvent and the anion. Therefore, the effects of substituents on the ring are diminished compared to the reaction in toluene, a nonpolar, non-hydrogen bonding solvent. The greater influence of the substituents in toluene is reflected in the larger value of ρ.

4. The accepted mechanism for acid-catalyzed hydrolysis of a carboxylic ester is:

$$1)\quad H^{+} + ArC(=O)OR \rightleftharpoons ArC(=\overset{+}{O}H)-OR$$

$$2)\quad ArC(=\overset{+}{O}H)-OR + H_2O \rightleftharpoons ArC(OH)(\overset{+}{O}H_2)(OR) \quad \left(\rightleftharpoons ArC(OH)(OH)(\overset{+}{O}H-R)\right)$$

$$3)\quad ArC(OH)(OH)(\overset{+}{O}HR) \rightleftharpoons ROH + ArC(=\overset{+}{O}H)-OH \quad \left(\rightleftharpoons ArC(=O)-OH + H^{+}\right)$$

If the second step in this reaction sequence is the rate limiting step, the overall reaction rate will be determined by the equilibrium constant for the first step and the rate constant for the second step. If the third step is rate limiting, the overall reaction rate will be determined by the equilibrium constants for the first and second steps and the rate constant for the third step.

The equilibrium constants for the first step in this sequence and the rate constants for the third step should have negative ρ values, and the rate and equilibrium constants for the second step should have positive ρ values. The very small overall ρ value presumably results from the effects of substituents in the first, and possibly the third, step effectively cancelling the effects of substituents in the second step.

CHAPTER 6

MIGRATIONS TO ELECTRON-DEFICIENT CENTERS

PROBLEM SOLUTIONS

1a. 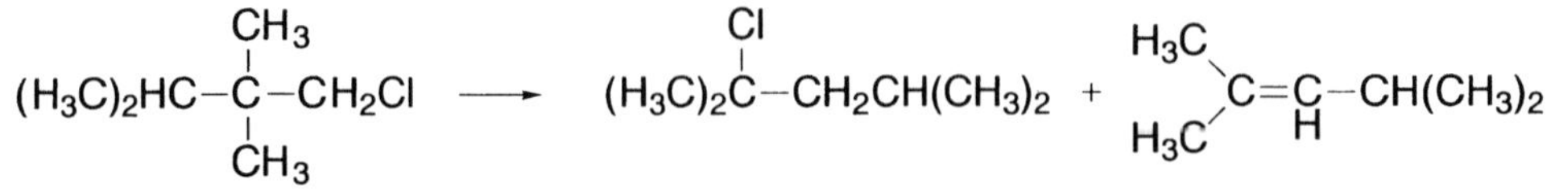 $(H_3C)_2HC{-}C(CH_3)_2{-}CH_2Cl \longrightarrow (H_3C)_2C(Cl){-}CH_2CH(CH_3)_2 + (H_3C)_2C{=}CH{-}CH(CH_3)_2$

1b. 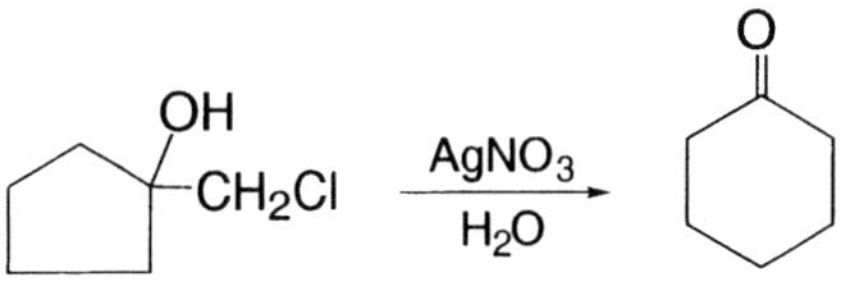 OH, CH_2Cl $\xrightarrow[H_2O]{AgNO_3}$ O

1c. 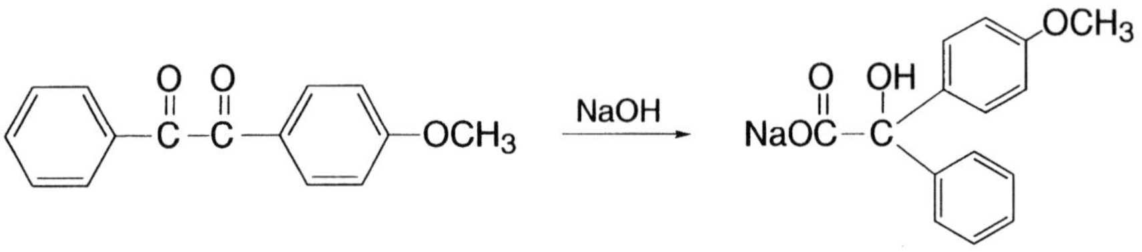 O O, C–C, OCH_3 $\xrightarrow{NaOH}$ OCH_3, O OH, NaOC–C

1d. 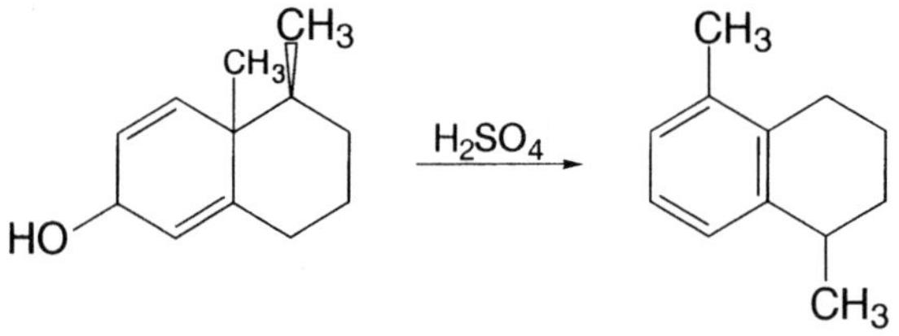 CH_3, CH_3, HO $\xrightarrow{H_2SO_4}$ CH_3, CH_3

1e. 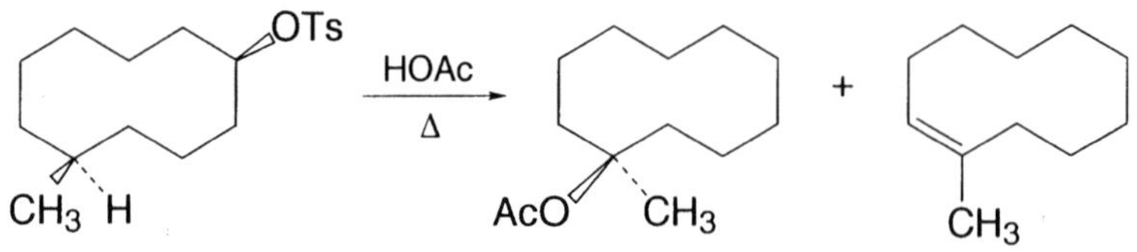 OTs, CH_3 H $\xrightarrow[\Delta]{HOAc}$ AcO CH_3 + CH_3

1f. 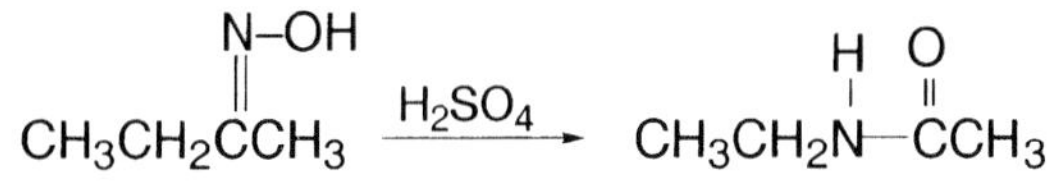 $CH_3CH_2C(=N{-}OH)CH_3 \xrightarrow{H_2SO_4} CH_3CH_2NH{-}C(=O)CH_3$

1g. 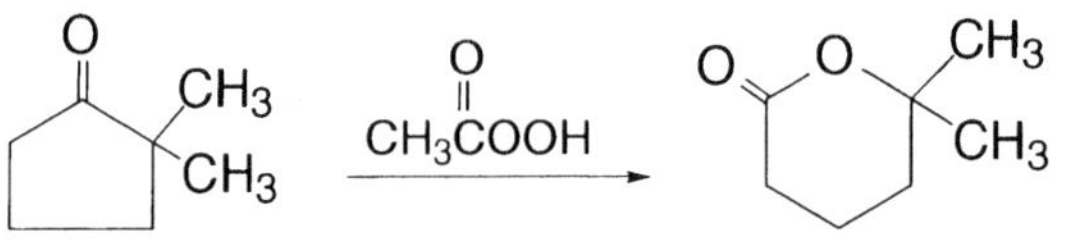 2,2-Dimethylcyclopentanone $\xrightarrow{CH_3C(=O)OOH}$ 6,6-dimethyl lactone (O, O, CH_3, CH_3)

2. As deuterotrifluoracetic acid attacks the double bond in this reaction, the newly forming cationic carbon interacts with a transannular hydrogen-carbon bond to form a hydrogen-bridged carbocation. Nucleophiles attack this carbocation from the direction *trans* to the bridging hydrogen atom, and *cis* to the deuterium atom.

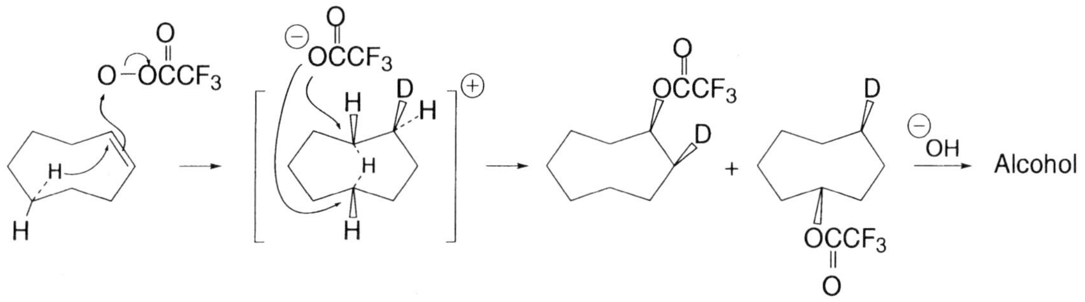

O–OCCF$_3$ (C=O), H, H → [$^{\ominus}$OCCF$_3$ (C=O), H, D, H, H, H]$^{\oplus}$ → OCCF$_3$ (C=O), D + D, OCCF$_3$ (C=O) $\xrightarrow{^{\ominus}OH}$ Alcohol

Of course, a rearranged product will also be formed during this reaction.

3a. $H_3CCH_2CH_2CH(H){-}CH_2{-}Cl$ + SbF_5 ⟶ $H_3CCH_2CH_2CH(H){-}CH_2{-}\overset{+}{Cl}{-}SbF_4$ (F^-) ⟶ $H_3CCH_2{-}CH_2{-}\overset{+}{C}HCH_3$

↓

$\overset{+}{C}H_2{-}C(H)(CH_3){-}CH_2CH_3$ ⟶ $(CH_3)_2\overset{+}{C}CH_2CH_3$ (F^-) ⟶ $(CH_3)_2C(F)CH_2CH_3$

3b. EtAlCl$_2$; O=C–CH$_3$ ⟶ H$_3$C–C(=$\overset{+}{O}$–$\overset{-}{Al}$EtCl$_2$) ⟶ H$_3$C, O–AlEtCl$_2$ (⊖), ⊕ ⟶ CH$_3$, $\overset{+}{O}$–$\overset{-}{Al}$EtCl$_2$

↓

CH$_3$, O + EtAlCl$_2$

3c. HSO$_3$O–H; CH$_3$; CH$_3$; CH$_3$ ⟶ HSO_4^-; H; CH$_3$; CH$_3$; ⊕; CH$_3$ ⟶ CH$_3$; CH$_3$; ⊕; CH$_3$

↘

$^-$OSO$_3$H; H; CH$_3$; ⊕; CH$_3$; CH$_3$ ⟶ CH$_3$; CH$_3$; CH$_3$

3d. $H{-}O{-}SO_2CF_3$ → → → $^{\ominus}OSO_2CF_3$ →

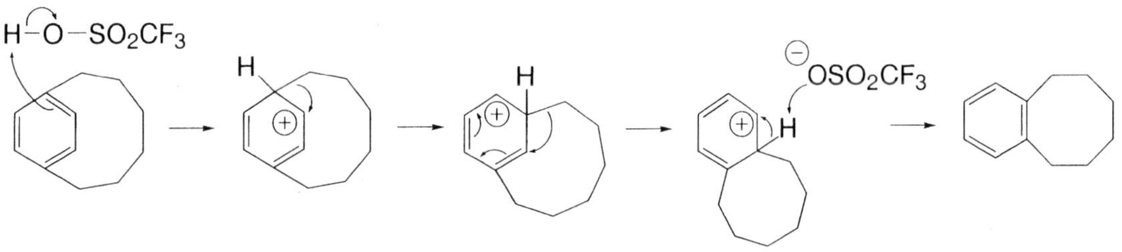

3e. $CH_3{-}MgBr$ → $BrMgO$, CH_3, $Br \frown MgBr_2$ → $BrMgO$, $CH_3^{\oplus}$, $^{\ominus}Br{-}MgBr_2$ → $BrMgO$, CH_3, CH_3 $\xrightarrow{(-MgBr)}$ CH_3, CH_3

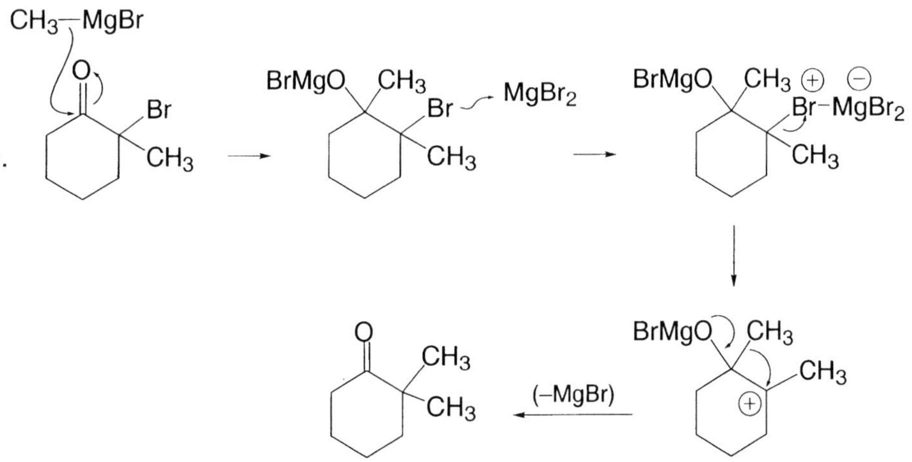

A mechanism involving direct displacement of the bromide ion by the Grignard reagent is very unlikely. Grignard agents are not effective in displacing halide ions, while they add readily to carbonyl groups. A carbocation rearrangement mechanism is supported by the fact that a second rearrangement product, 1-methyl-1-acetylcyclopentane, is also formed in the reaction.

3f.

H_3C $\ddot{O}H$ H^{+} → H_3C $\overset{+}{O}H_2$ → CH_3 (+) H–O–OH + H_2O

↓

OH_2 O (+) CH_3 ← H_3C H–O–$\overset{+}{O}H_2$ ← H_3C H–$\overset{+}{O}$–OH

↓

$\overset{+}{O}H_2$ O CH_3 → H–$\overset{+}{O}$ $\ddot{O}H$ CH_3 → OH H–$\overset{+}{O}$ CH_2CH_2-C–CH_3

↓

OH O CH_2CH_2-C–CH_3

3g. $\overset{\oplus}{C}(C_6H_5)_3$ H H ⟶ $(C_6H_5)_3CH$ + ⊕

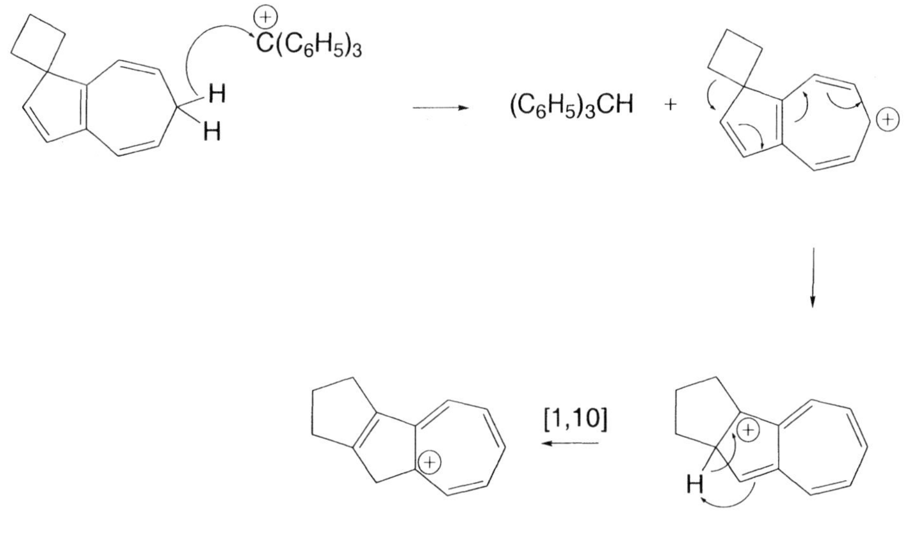

⟶ [1,10] ⟵ ⊕ H

3h. O_2N O CH_3 CH_3 $\xrightarrow{H_2SO_4}$ O_2N H O⊕ CH_3 CH_3 ⟶ O_2N HÖ $\overset{\oplus}{C}(CH_3)_2$

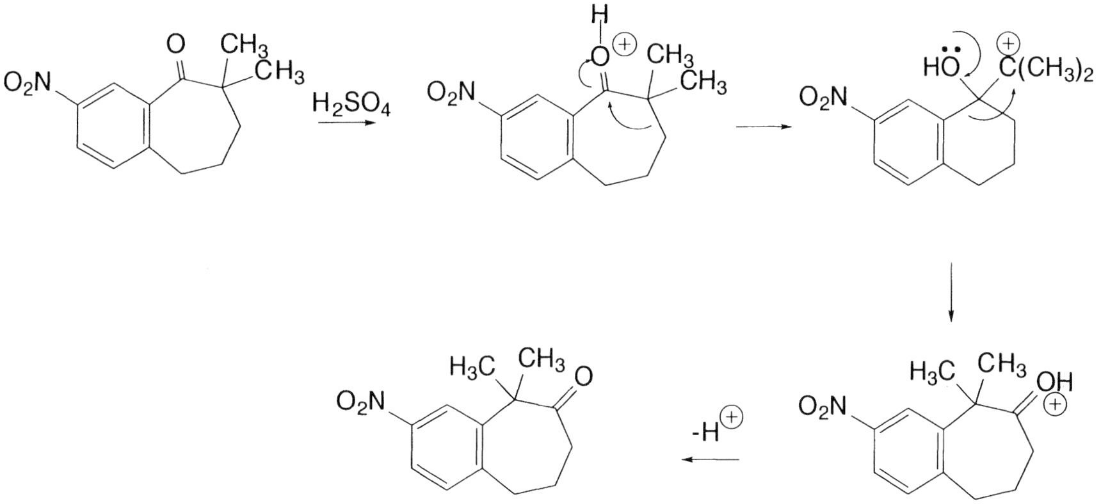

⟶ O_2N H_3C CH_3 OH⊕ $\xleftarrow{-H^{\oplus}}$ O_2N H_3C CH_3 O

3i.

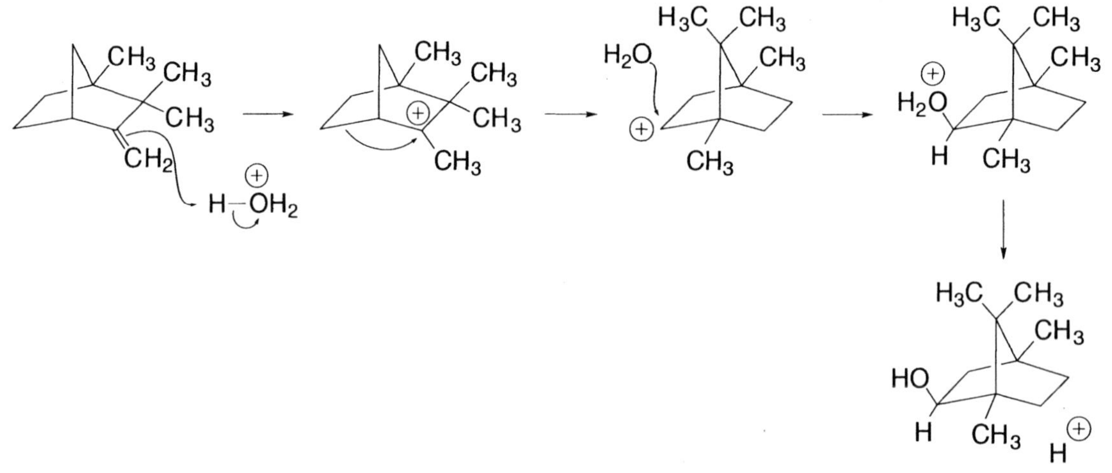

CH_3 CH_3 CH_3 CH_2 $H-\overset{+}{O}H_2$ → CH_3 CH_3 CH_3 CH_3 → H_3C CH_3 H_2O CH_3 CH_3 → H_3C CH_3 CH_3 $H_2\overset{+}{O}$ H CH_3 → H_3C CH_3 CH_3 HO H CH_3 H^+

3j.

CH_3 O CH_3 $H-OSO_3H$ → CH_3 $H-\overset{+}{O}$ CH_3 → CH_3 HO H CH_3 → CH_3 OH_2 HO $(CH_2)_3\overset{+}{C}HCH_3$ → (2 steps) → CH_3 OH HO $(CH_2)_3CHCH_3$

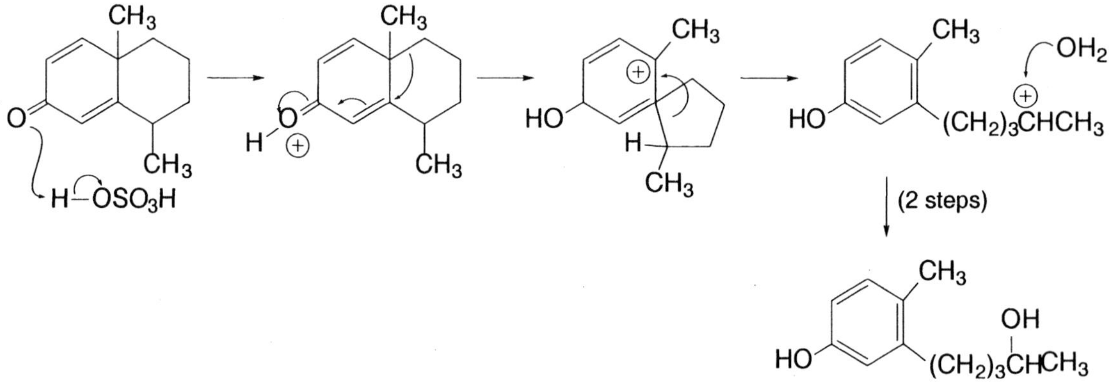

3k.

HO CH_3 CH_2CH OCH_3 OCH_3 $SnCl_4$ → HO CH_3 CH_2-CH $\overset{+}{O}CH_3$ → OH CH_3 OCH_3 → $\overset{+}{O}-H$ CH_3 $-OCH_3$ → $-H^+$ → O CH_3 $-OCH_3$

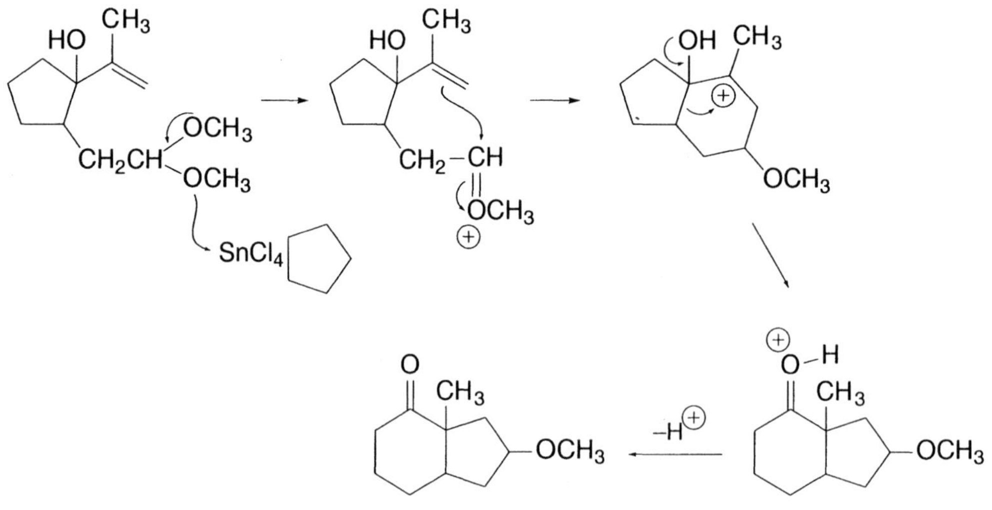

31.

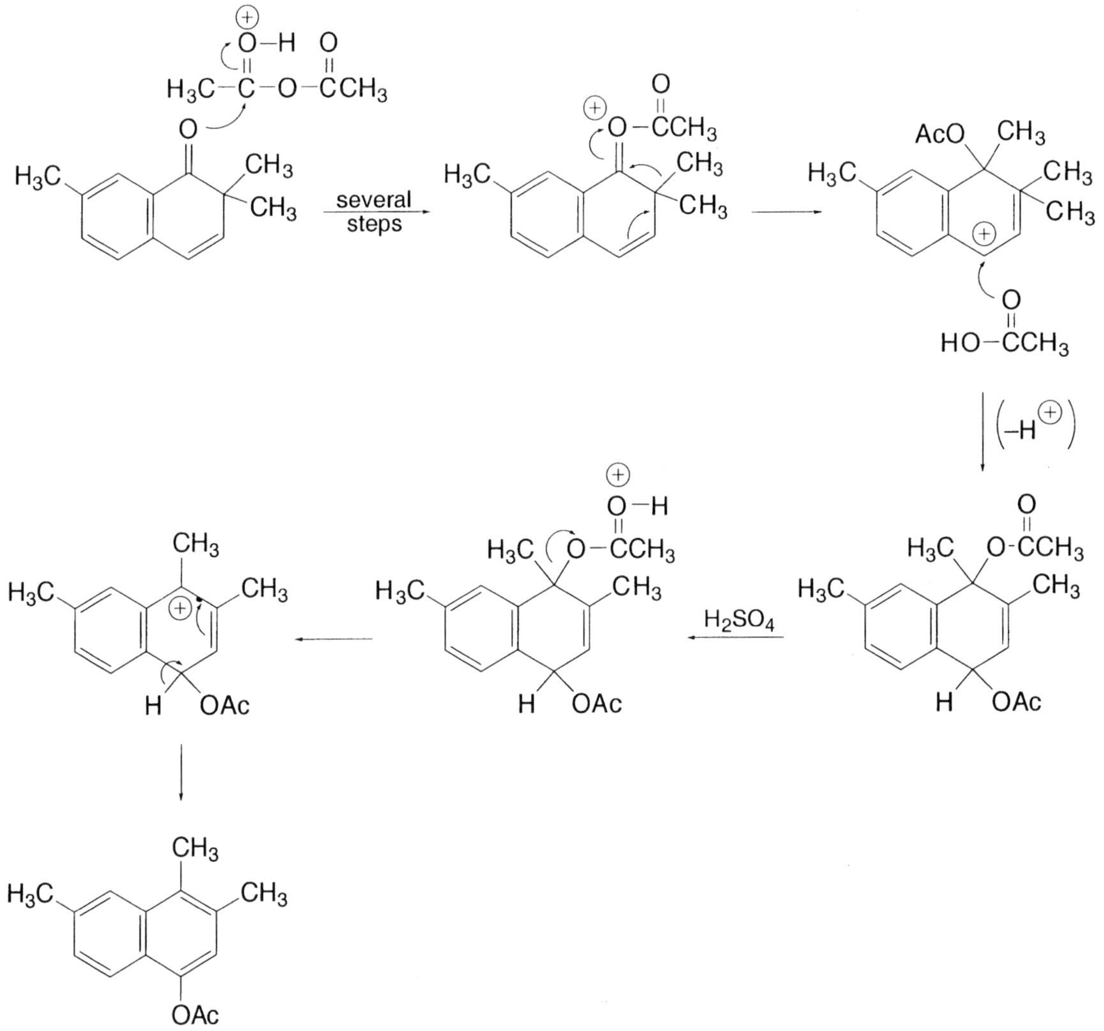

several steps

$(-H^{\oplus})$

H_2SO_4

3m.

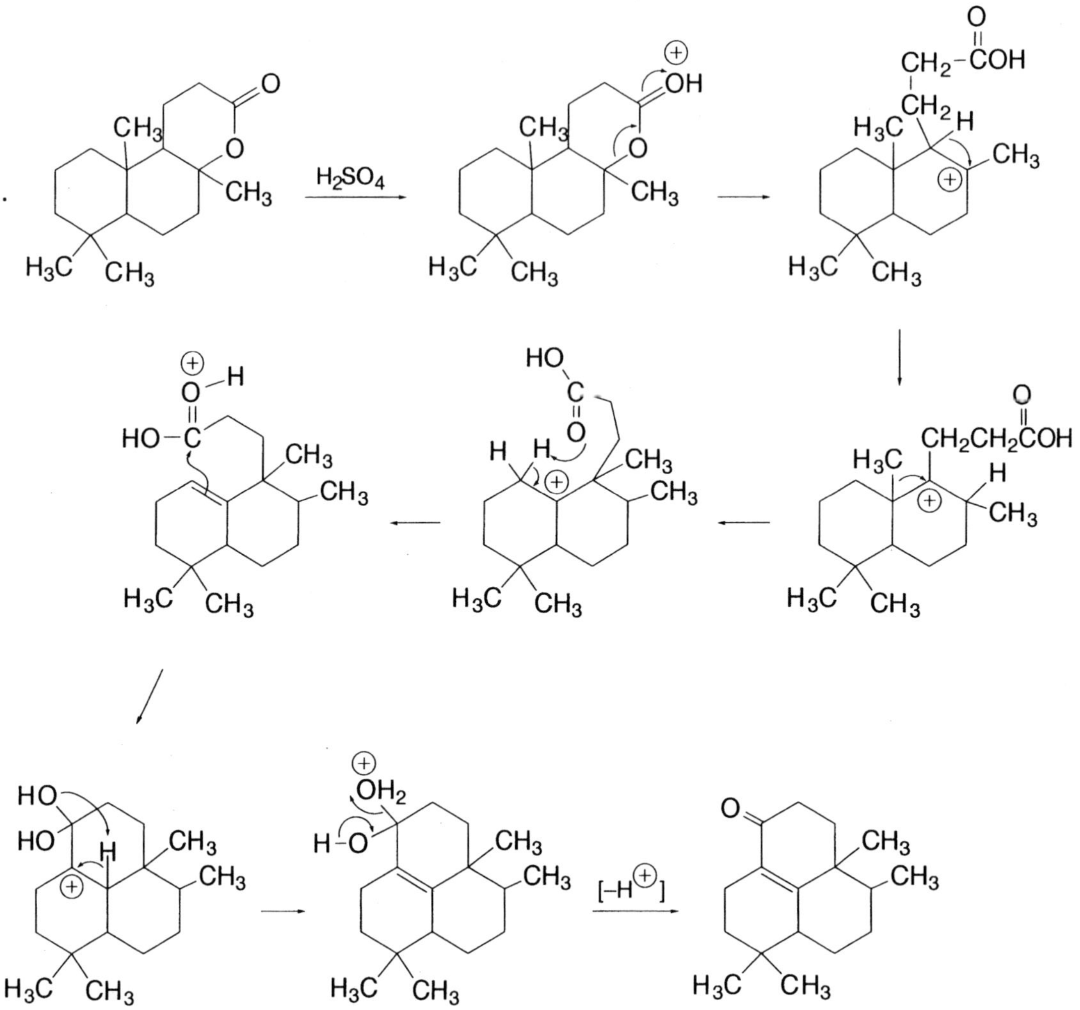

H_2SO_4

$[-H^{\oplus}]$

3n.

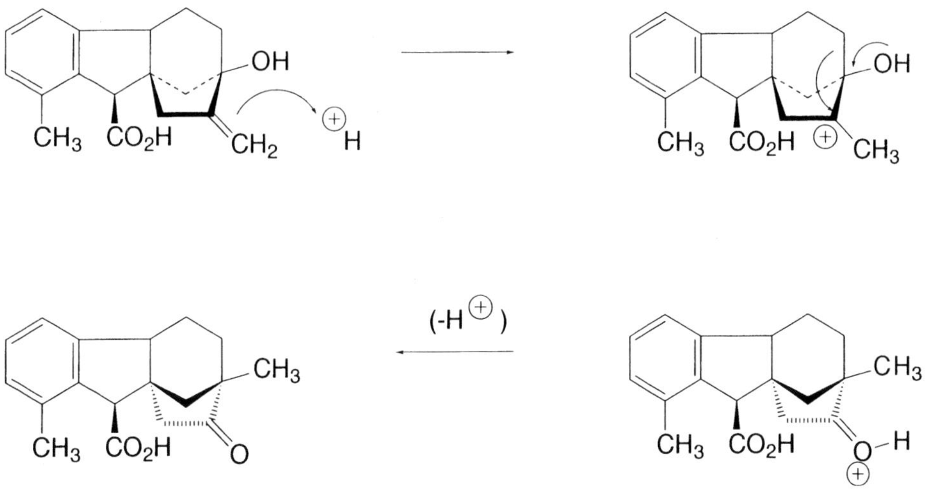

CH_3 CO_2H CH_2 OH ⊕H → CH_3 CO_2H ⊕ CH_3 OH

(-H⊕)

CH_3 CO_2H CH_3 O ← CH_3 CO_2H CH_3 O-H ⊕

Note: Tertiary alcohols usually react more rapidly with acids than do double bonds. However, in this molecule protonation of the hydroxy group and loss of water would form a nonplanar, high energy bridghead carbocation. Thus a carbocation is more easily formed by protonation of the double bond.

3o.

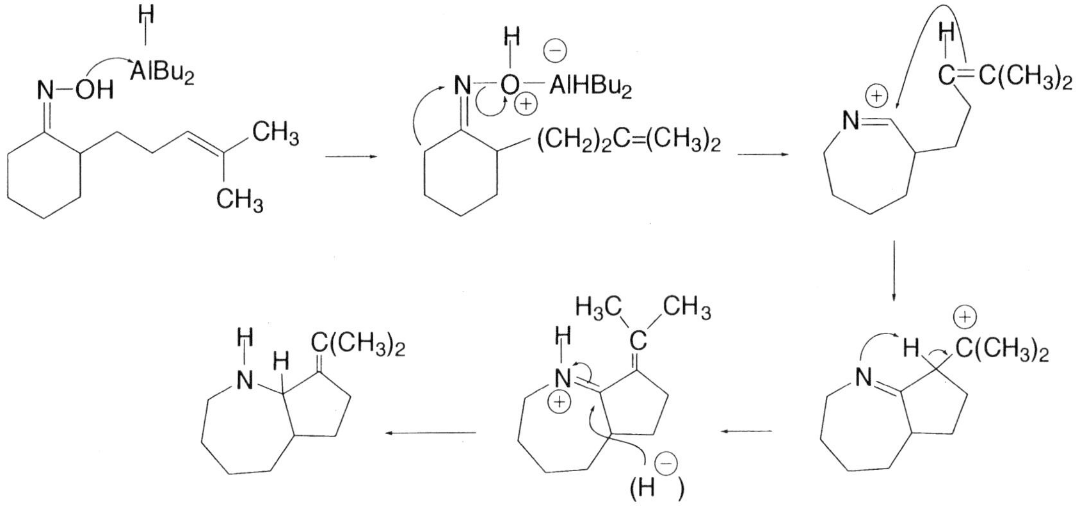

N-OH H $AlBu_2$ CH_3 CH_3 → H ⊖ N-O⊕-$AlHBu_2$ $(CH_2)_2C{=}(CH_3)_2$ → ⊕ N= H C=$C(CH_3)_2$

H N H $C(CH_3)_2$ ← H_3C CH_3 C H N⊕ ($H^⊖$) ← ⊕ N= H $C(CH_3)_2$

3p.

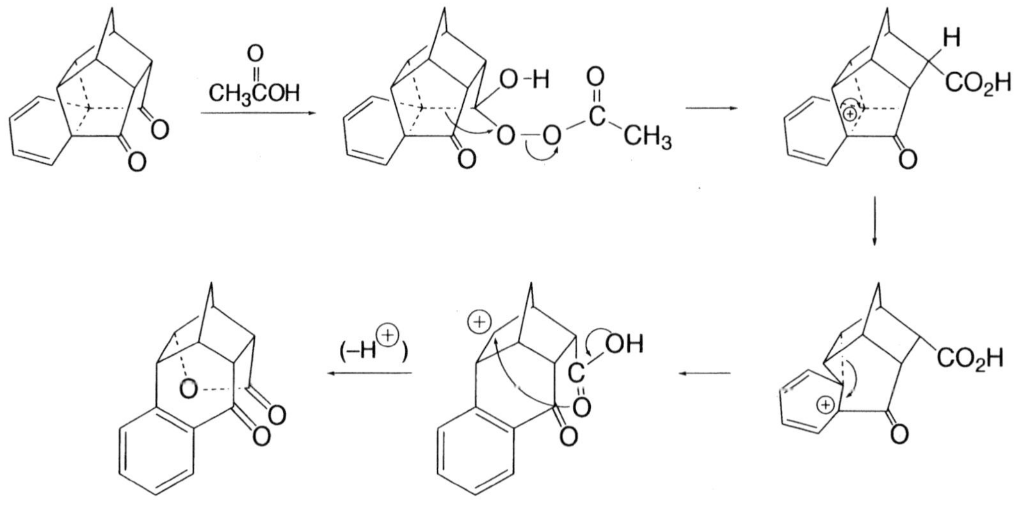

A difficult mechanism to visualize in this complex polycyclic system. Congratulations if you worked it out!

Note: migration of the acyl group rather than a ring carbon would also account for formation of the product.

CHAPTER 7

NEIGHBORING GROUP EFFECTS AND "NONCLASSICAL" CATIONS

PROBLEM SOLUTIONS

1a. TsO ... N ... O $\xrightarrow{\text{LiCl}}$ $ClCH_2$ ... N ... O

1b. CH_3, I, OTs $\xrightarrow{\text{HOAc}}$ CH_3, I, OAc + CH_3, OAc, I

1c. CH_3 H, CH_3, OH, H, H $\xrightarrow{\text{HCl}}$ CH_3 H, CH_3, Cl, H, H + H, Cl, CH_3, H, H, CH_3

1d. OTs, H, C, C, H, CH_3, C_2H_5 $\xrightarrow{\text{HOAc}}$ C_2H_5, H, H, C, C, CH_3, OAc + H, CH_3, C, C, C_2H_5, AcO, H

2a.

2b.

2c.

If a "nonclassical", bridged carbocation were the only intermediate formed in the second step of this reaction, nucleophilic attack on the cation would have to occur from the side of the three membered ring opposite to the bond bearing the carboxy substituent. The fact that the carboxy group acts as the nucleophile can be most easily explained by formation of an unbridged carbocation, as shown above. However, the evidence does not eliminate the possible formation of a bridged intermediate which opens to the "classical" carbocation, or which initially forms a dibromide which is then converted to the lactone.

2d. $Cl{-}CH_2C({=}S){-}OR \rightarrow$ [cyclic S^+ intermediate, Cl^-, attack by OH_2] $\rightarrow$ [$CH_2{-}C(OR)(OH^+){-}S$ ring] $\rightarrow$ [$CH_2{-}C(OR)(OH){-}S^+H$ ring] $\rightarrow$ $HSCH_2{-}C({=}OH^+){-}OR$ Cl^- $\rightarrow$ $HSCH_2COR$ (C=O) + HCl

2e. [1-methyl-3-(2-bromo-2-methylpropyl)indole, CH_2, Br, $(CH_3)_2$, CH_3] $\rightarrow$ [spirocyclopropane indoleninium ion, Br^-, CH_3, CH_3] $\rightarrow$ [1-methyl-3-($C(CH_3)_2CH_2Br$)indole]

3. It seems probable that water, ethanol, and acetic acid all initially add to the alkylated carbonyl group of **6**. The addition of water would form an intermediate which could rapidly open to an ester, while the addition of ethanol would be essentially irreversible under basic or neutral conditions. In contrast, the addition of acetic acid would b reversible, reforming **6**, because acetic acid is a better leaving group than water or ethanol and a better ionizing solvent than ethanol. The reaction of acetic acid at a saturated carbon of **6**, though slower than its addition to the carbonyl group, would thus yield the more stable product - the only one actually detected.

4a. A curved Hammett plot for the solvolysis of **19** most likely results from a change in mechanism, from formation of a classical (benzylic) cation when X is strongly electron-donating to a "nonclassical" cation, resulting from participation by the double bond, when X is less strongly electron-donating.

4b. Regardless of whether **19** yields a classical or nonclassical ion, its ionization should always be assisted by electron-donating substituents and inhibited by electron-withdrawing substituents. Thus, the slope of the line (ρ) should always be negative.

4c. The effects of substituents would be greatest when **19** is converted to a classical benzylic cation. Thus, the slope of the line should have the largest (negative) value when substituents are strongly electron-donating.

CHAPTER 8

REARRANGEMENTS OF CARBANIONS AND FREE RADICALS

PROBLEM SOLUTIONS

1a. C_6H_5–CH_2–O–CH_2–C_6H_5 $\xrightarrow{CH_3Li}$ C_6H_5–CH_2–CH(C_6H_5)–OLi

1b. 3-(bromomethyl)cyclopentene (CH_2Br) $\xrightarrow[ROOR]{Bu_3SnH}$ cyclohexene

1c. $C_6H_5CH{=}CH{-}CH_2{-}\overset{\oplus}{S}(CH_3)_2$ $\xrightarrow{BuLi}$ $C_6H_5{-}CH(CH_2{-}S{-}CH_3){-}CH{=}CH_2$

1d. C_6H_5–CH_2–$\overset{\oplus}{N}(CH_2C_6H_5)(CH_3)$–$CH_3$ $\xrightarrow[liquid\ NH_3]{NaNH_2}$ 2-CH_3-C_6H_4–CH(C_6H_5)–$N(CH_3)_2$

1e. [1-bromo-1,3-dihydrobenzo[c]thiophene 2,2-dioxide] $\xrightarrow{NaOCH_3}$ [benzocyclobutene]

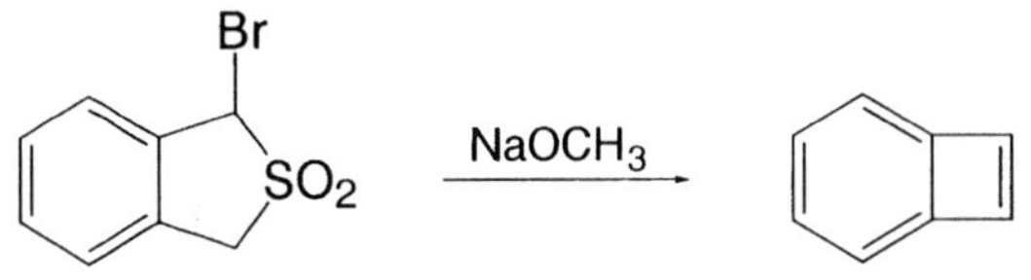

1f. $C_6H_5CH{=}CH{-}CH(C_6H_5){-}CHBrCH_3$ $\xrightarrow[2)\ CH_3CHO]{1)\ Li}$ $C_6H_5CH{=}CH{-}CH(CH_3){-}CH(CH(CH_3)OLi){-}CHC_6H_5$ or

$C_6H_5CH{=}CH{-}CH(CH(CH_3)OLi){-}CH(C_6H_5)CH_3$

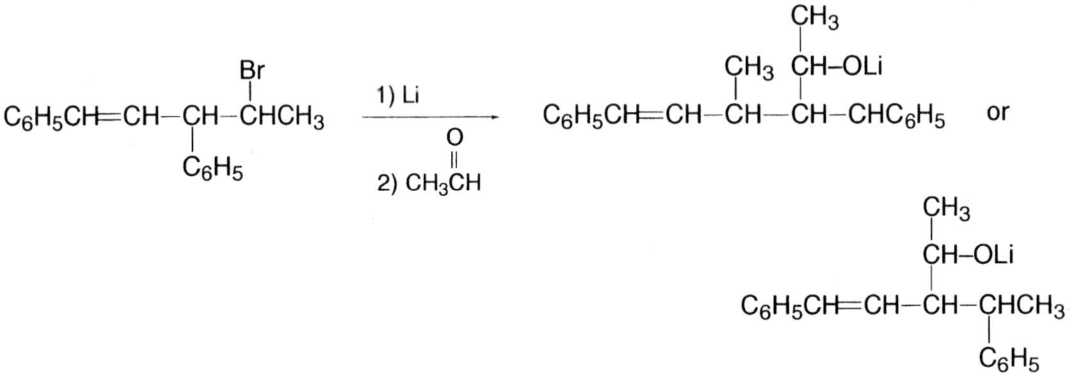

Migration of the $C_6H_5CH{=}CH-$ group should occur more rapidly than migration of the phenyl group, but migration of the phenyl group would yield a more stable carbanion.

1g. [1-methyl-1-(CH_2CHO)-indane, with H_3C and $CH_2\overset{O}{\overset{\|}{C}}H$ substituents] $\xrightarrow[\Delta]{ROOR}$ [2-methyltetralin, with H and CH_3]

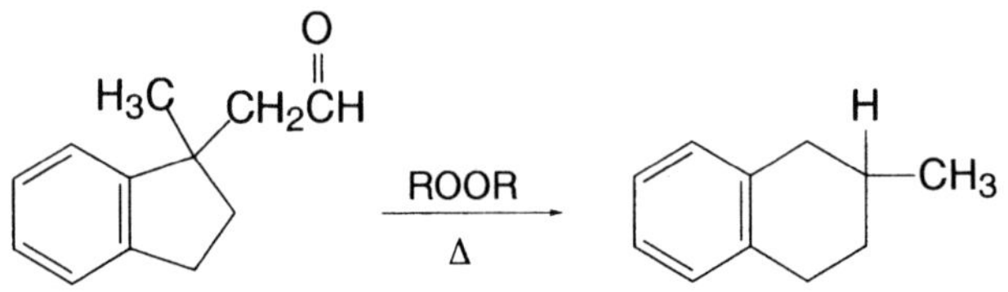

1h. [cyclic transition state: $CH_2{=}CH{-}CH(CH_3)$ / H, $O{-}\overset{\ominus}{C}(CH_3){-}C{\equiv}N$] $\longrightarrow$ $H_3C\overset{O}{\overset{\|}{C}}{-}CH(CH_3){-}CH{=}CH_2$ + $^{\ominus}CN$

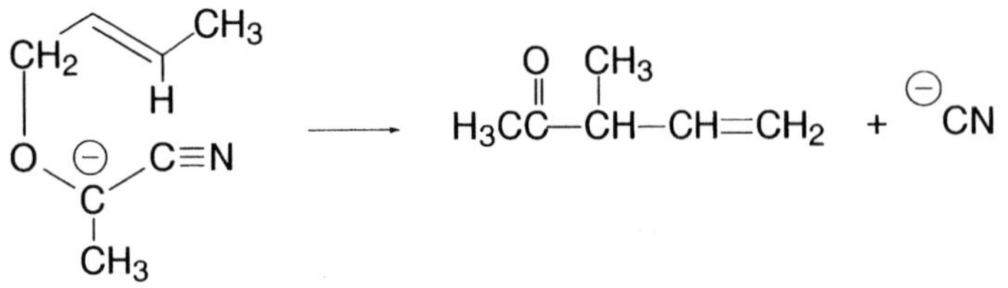

2a.

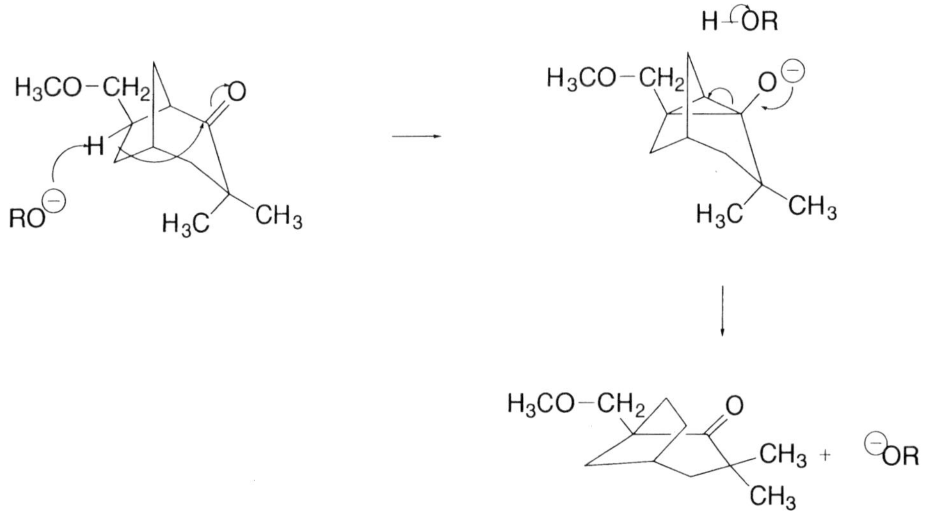

We can reasonably conclude that abstraction of the proton is concerted with formation of the three membered ring, placing the charge principally on the oxygen atom rather on carbon (If an appreciable charge had developed on the carbon atom, elimination of the methoxide group, as shown below, would have been expected.)

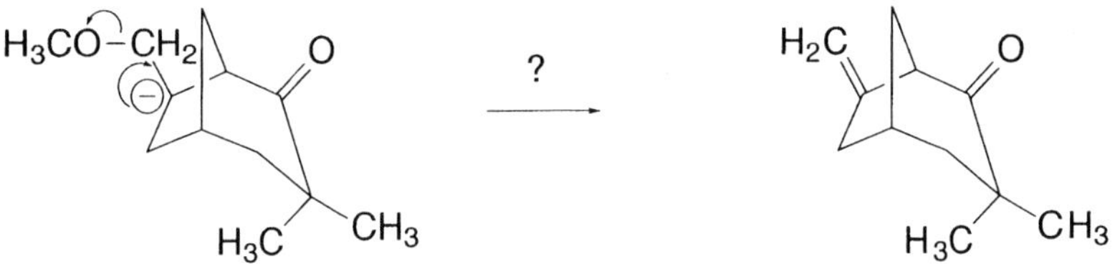

Since the protonation step is likely to proceed via a transition state similar to that of the deprotonation step (although at a different carbon), little charge should be present on carbon in that step either.

2b.

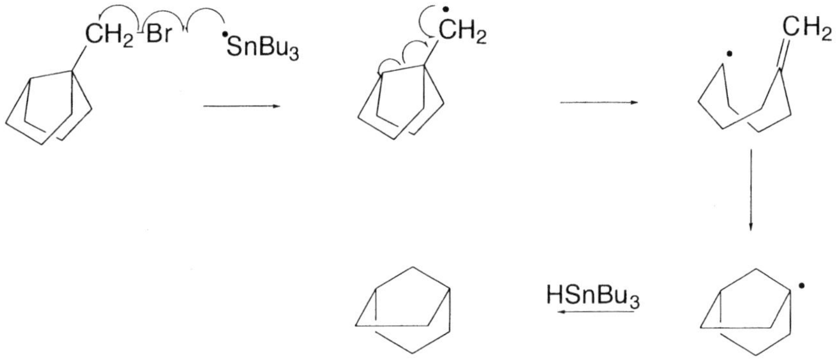

2c. (Z)-$CH_3CH{=}CHCH_2CH_2CHO$ + $\cdot OR$ ⟶ $CH_3CH{=}CHCH_2CH_2\dot{C}{=}O$ ⟶ $CH_3CH{=}CHCH_2\dot{C}H_2$ + CO

⇌ $CH_3\dot{C}H{-}CH(CH_2)_2$ (cyclopropyl) ⇌ (E)-$CH_3CH{=}CHCH_2\dot{C}H_2$ + $H{-}\overset{O}{\overset{\|}{C}}C_5H_9$ ⟶ (E)-$CH_3CH{=}CHCH_2CH_3$ + $\cdot\overset{O}{\overset{\|}{C}}C_5H_9$

2d. $CH_2(CO_2CH_3)_2$ $\xrightarrow{NaOCH_3}$ $(CH_3O_2C)_2CH^{\ominus}$ + $CH_2{=}CH{-}\overset{O}{\overset{\|}{C}}{-}C(Br)(CH_3)_2$

⟶ $(CH_3O_2C)_2CHCH_2\overset{\ominus}{C}H{-}\overset{O}{\overset{\|}{C}}{-}C(Br)(CH_3)_2$

⟶ $(CH_3O_2C)_2CH{-}CH$ (cyclopropanone, with $C{=}O$ and $C(CH_3)_2$) + $CH_3O^{\ominus}$

⟶ $(CH_3O_2C)_2CHCH_2CH$ (cyclopropane ring: $C(OCH_3)(O^{\ominus})$, $C(CH_3)_2$)

$\xrightarrow{CH_3OH}$ $(CH_3O_2C)_2CHCH_2CH_2{-}C(CH_3)_2{-}\overset{O}{\overset{\|}{C}}OCH_3$

2e.

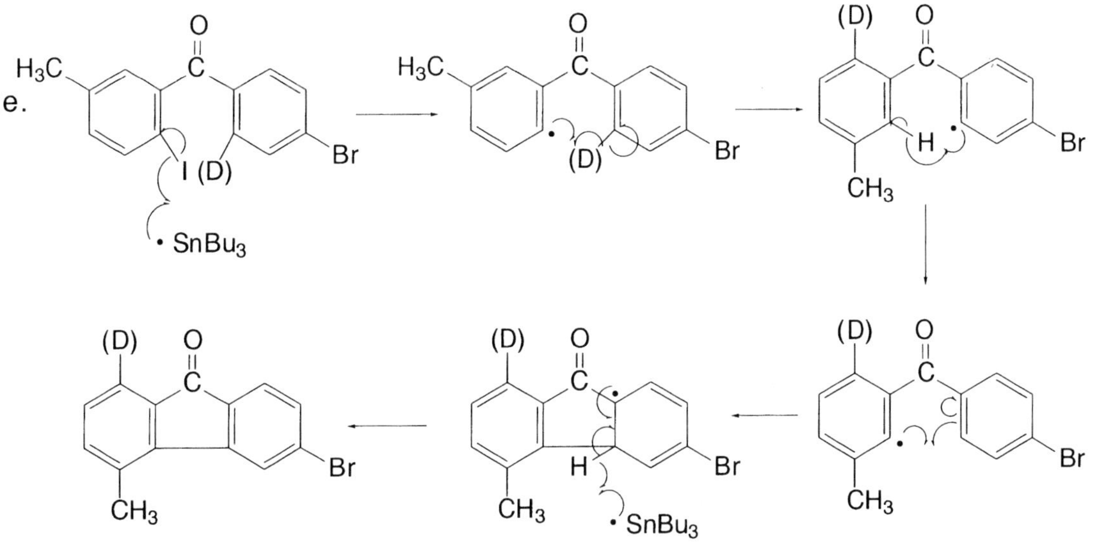

2f.

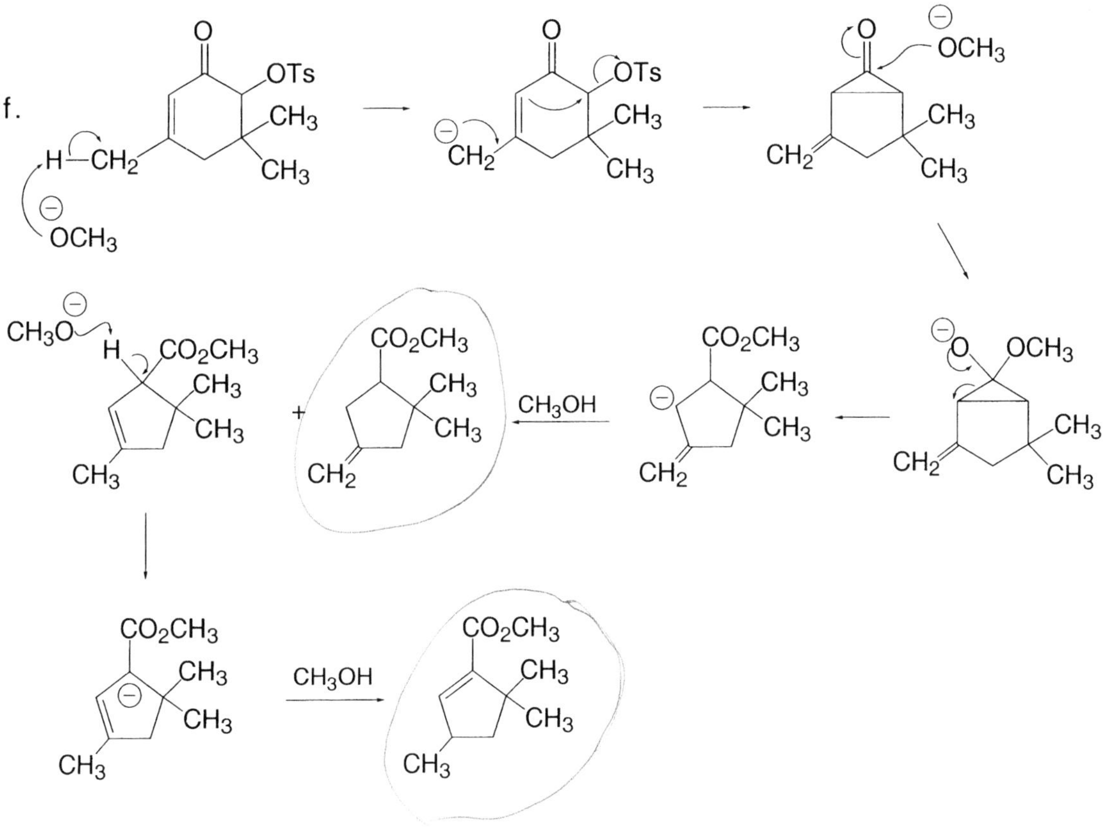

and:

$CH_3O^{\ominus}$ CH_3OH CH_3OH $^{\ominus}OCH_3$

2g. $Cl_3C\cdot$ + $CH_2{=}CH{-}C(CH_3)_2{-}CCl_2(C{\equiv}N)$ → $Cl_3CCH_2{-}\dot{C}H{-}C(CH_3)_2{-}CCl_2(C{\equiv}N)$ → Cl_3CCH_2CH (cyclic, N•, C, CCl_2, C, CH_3, CH_3) → $Cl_3CCH_2{-}CH(C{\equiv}N){-}C(CH_3)_2{-}\dot{C}Cl_2$ + $Cl{-}CCl_3$ → $Cl_3CCH_2{-}CH(C{\equiv}N){-}C(CH_3)_2{-}CCl_3$ + $\cdot CCl_3$

2h. This reaction requires three [1,6] (antarafacial) hydrogen migrations followed by an electrocyclic reaction.

CH_2 HC $(CH_2)_2-CH=CH_2$ [1,6] CH_3 CH H $C-CH_2-CH=CH_2$ H [1,6] C_2H_5 CH H $CH=CH_2$ CH [1,6] C_3H_7 CH_2 CH $CH_3CH_2CH_2$

2i.

H H

2j.

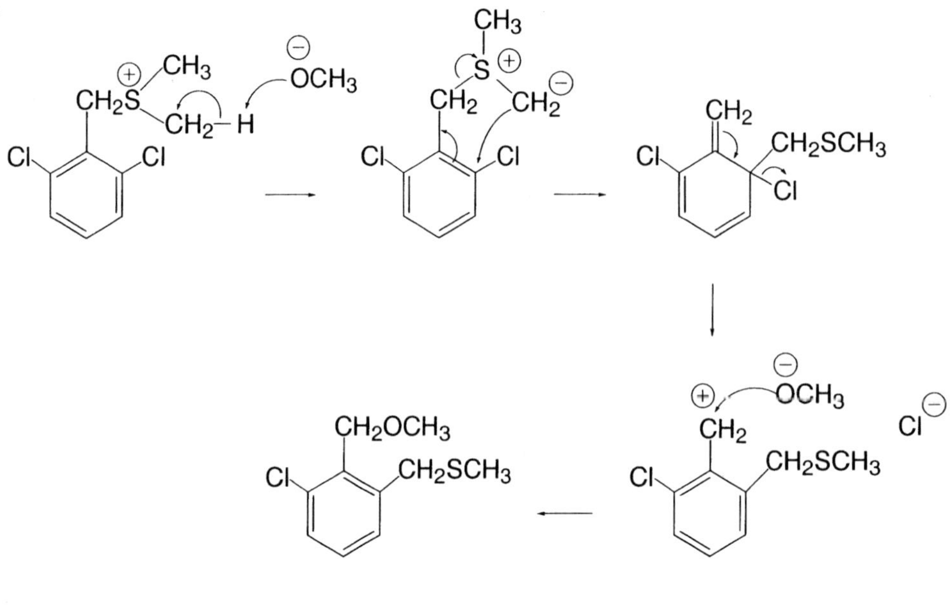

CH₂S⁺(CH₃)CH₂–H ⁻OCH₃ Cl Cl

2k. One possible mechanism:

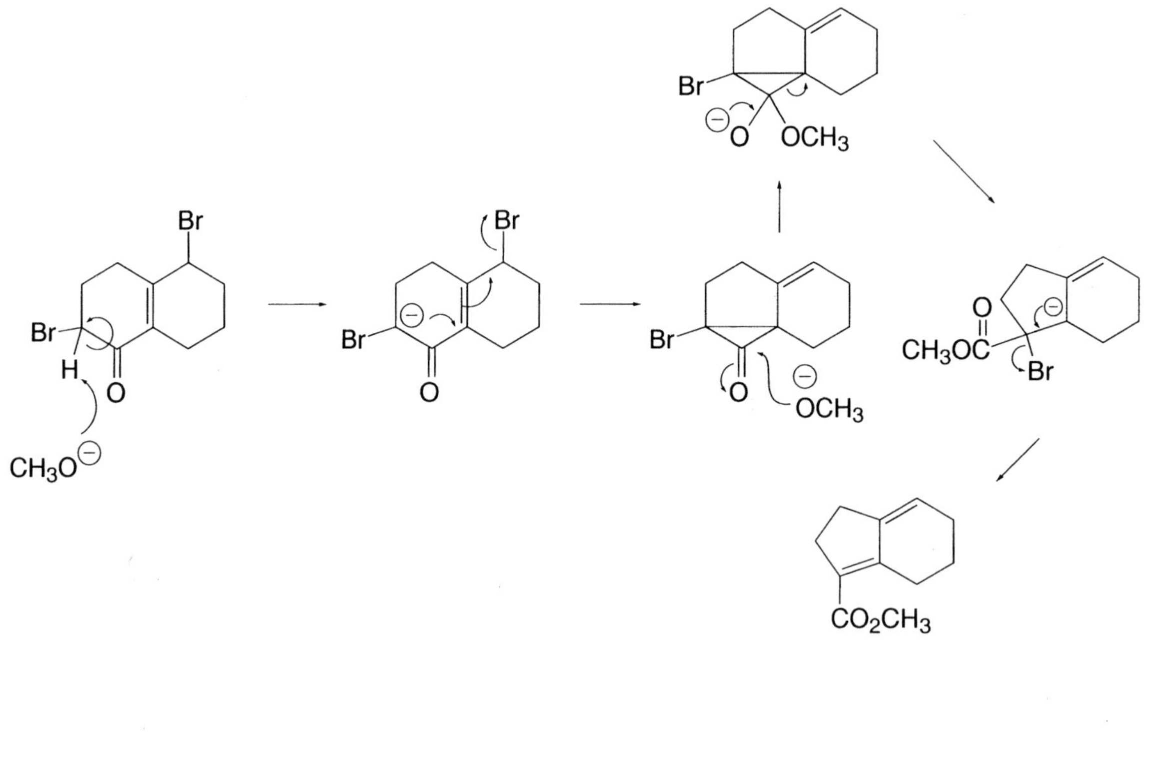

Br Br H O CH₃O⁻ OCH₃ CH₃OC CO₂CH₃

A second possible mechanism:

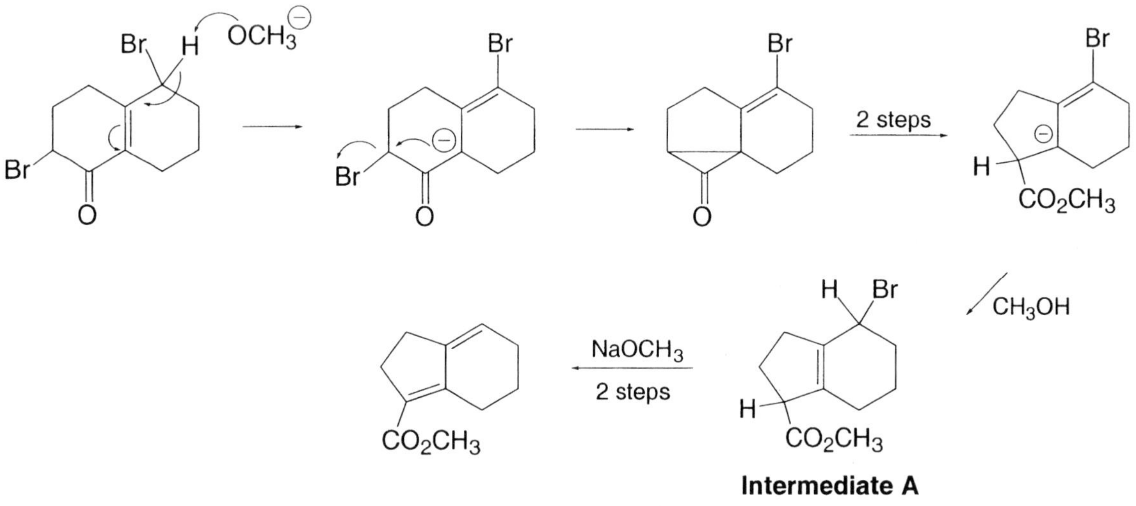

2 steps

CH_3OH

NaOCH$_3$ 2 steps

Intermediate A

A third possible mechanism:

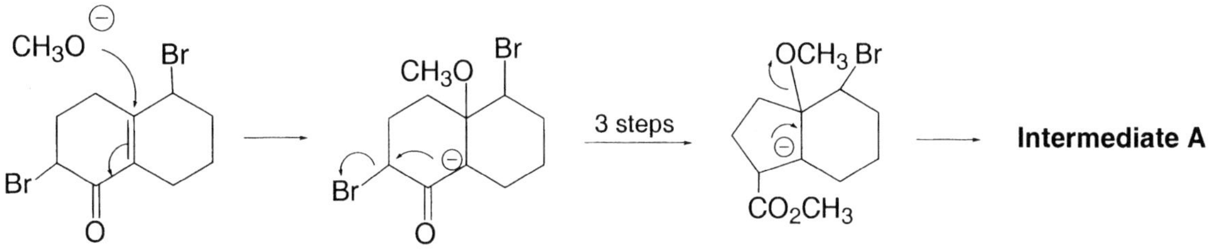

3 steps

Intermediate A

21.

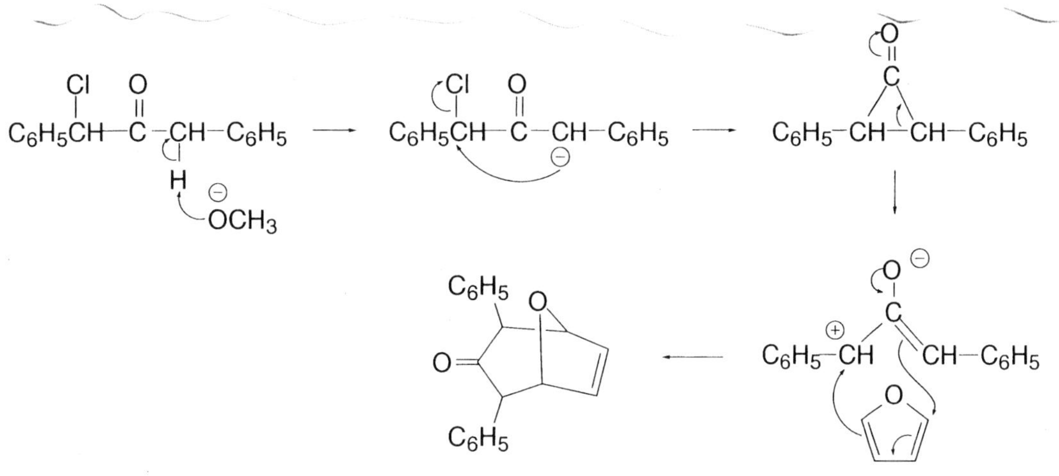

2m.

O

O•

O

$(CH_2)_3CH_2\cdot$

2n.

N

SO_2R

CH_2

N

SO_2R

N

SO_2R

N

SO_2R

H—$SnBu_3$

2o.

R

R

H_2C

CH_2

R

CH_2

R

CH_2

Bu_3SnH

R

H

CHAPTER 9

CARBENES, CARBENOIDS, AND NITRENES

PROBLEM SOLUTIONS

1. The relative stabilities of the singlet forms of these 7-norbornenyl carbenes suggest that they resemble 7-norbornenyl cations, and may similarly show "homoaromatic" character, as illustrated below for caebene c.

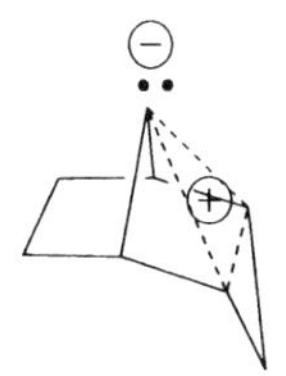

As with the carbocations, interaction of the empty orbital with a cyclopropane ring provides greater stabilization than with a π bond, which in turn provides much greater stabilization than interaction with a σ bond.

2. The first reaction in this problem requires a hydrogen migration in a carbene intermediate while the thiomethyl group migration in the second reaction presumably proceeds by neighboring group interaction of the sulfur atom with the carbene carbon.

$$C_6H_5\ddot{C}-CH(H)OCH_3 \longrightarrow C_6H_5CH{=}CHOCH_3$$

$$C_6H_5\ddot{C}-CH_2SCH_3 \longrightarrow C_6H_5-\overset{\ominus}{C}-CH\ (\text{bridged by } \overset{\oplus}{S}-CH_3) \longrightarrow C_6H_5-C(SCH_3){=}CH_2$$

Formation of the three-membered ring in the second reaction is more likely than in the first reaction because sulfur atoms are much better nucleophiles than oxygen atoms, and because a three-membered ring containing a sulfur atom is less strained than a similar ring containing an oxygen atom (see Chap. 7).

3a. C≡C−CHN₂ / CH₂CH=CH₂ —Δ→ C≡C−C̈H / CH₂CH=CH₂ ⟷ C̈−C≡CH / CH₂CH=CH₂ → H−C≡C

3b. O, Cl, Cl, Zn → O → H, H, O → O, H

3c. ⊖OR, H, CH₂CH₃, Br−C=C, (CH₂)₃CH₃ → ⊖, CH₂CH₃, Br−C̈=C, (CH₂)₃CH₃ → CH₂CH₃, :C=C, H, CH₂, CH−CH₂, H₃C → CH₂CH₃, H₃C

3d. $CHCl_3$ + KOH → $Cl_2\ddot{C}$ $O^{\ominus}$ → H $Cl_2C^{\ominus}$ O —H_2O→ $HO^{\ominus}$ H O Cl_2CH

→ Cl CH Cl $O^{\ominus}$ → $HO^{\ominus}$ H C Cl $O^{\ominus}$ → O H CH Cl $O^{\ominus}$ → $O^{\ominus}$ CH Cl OH

→ O HC OH —$OH^{\ominus}$→ O HC $O^{\ominus}$

3e. CH_3 $O^{\ominus}$ $\ddot{C}Cl_2$ CH_3 → CH_3 O $^{\ominus}CCl_2$ CH_3 —H_2O→ CH_3 O $CHCl_2$ CH_3

3f.

O, N^{+}, N^{-}, O

Rh(OAc)2

O, O, $:Br-CH_2-C_6H_5$

O, Br^{+}, $CH_2-C_6H_5$, O^{-}

O, Br, $O-CH_2\ C_6H_5$

3g.

13, C≡C–C≡C

13, :C=C, C≡C–C_6H_5, C_6H_5

C≡C–$\overset{13}{C}$≡C

or:

13, C≡C–C≡C

C_6H_5, 13, C=C:, C≡C, C_6H_5

C≡C–$\overset{13}{C}$≡C

3h. $(CH_3O)_2C:$ C≡C–CO_2CH_3 (CO_2CH_3)

$(CH_3O)_2C^{+}$–C=C^{-}–CO_2CH_3 (CO_2CH_3)

O=C–C≡C–CO_2CH_3 (OCH_3)

CH_3O_2C, CO_2CH_3, C=C, OCH_3, $(CH_3O)_2C^{+}$, C, O^{-}, C≡C–$\overset{O}{\overset{\|}{C}}OCH_3$

$CH_3O\overset{O}{\overset{\|}{C}}$, $\overset{O}{\overset{\|}{C}}OCH_3$, CH_3O, OCH_3, CH_3O, O, C≡C–CO_2CH_3

3i.

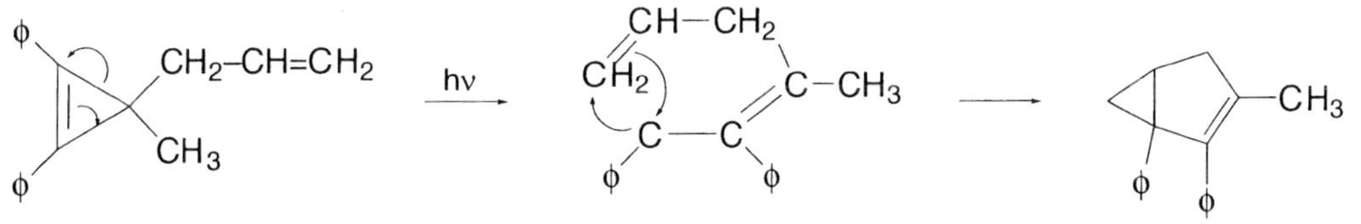

ϕ ϕ CH_2–CH=CH_2 CH_3 → hν → CH–CH_2 CH_2 C–CH_3 C C ϕ ϕ → CH_3 ϕ ϕ

3j.

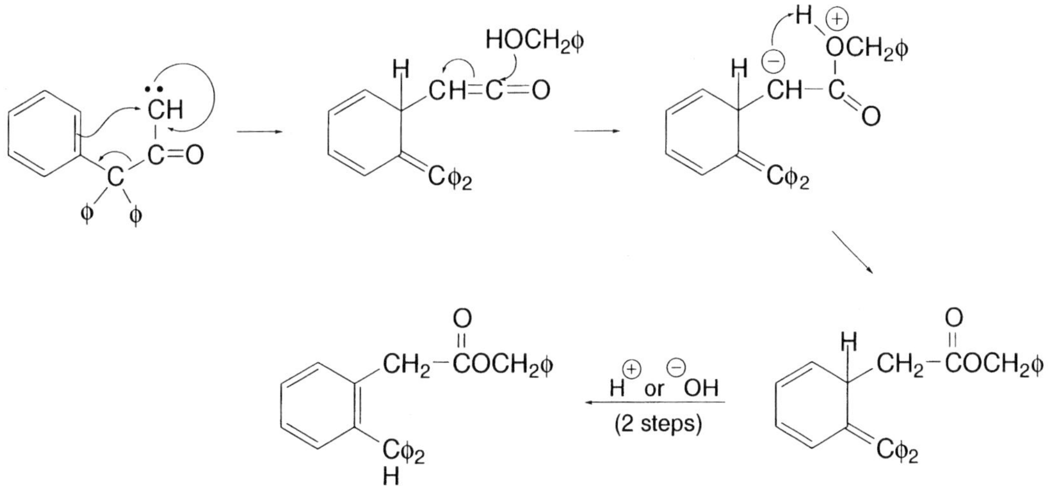

CH C=O C ϕ ϕ → $HOCH_2\phi$ H CH=C=O $C\phi_2$ → H ⊕ $OCH_2\phi$ H ⊖ CH–C=O $C\phi_2$ →

H O CH_2–$COCH_2\phi$ $C\phi_2$ → H⊕ or ⊖OH (2 steps) → O CH_2–$COCH_2\phi$ $C\phi_2$ H

3k.

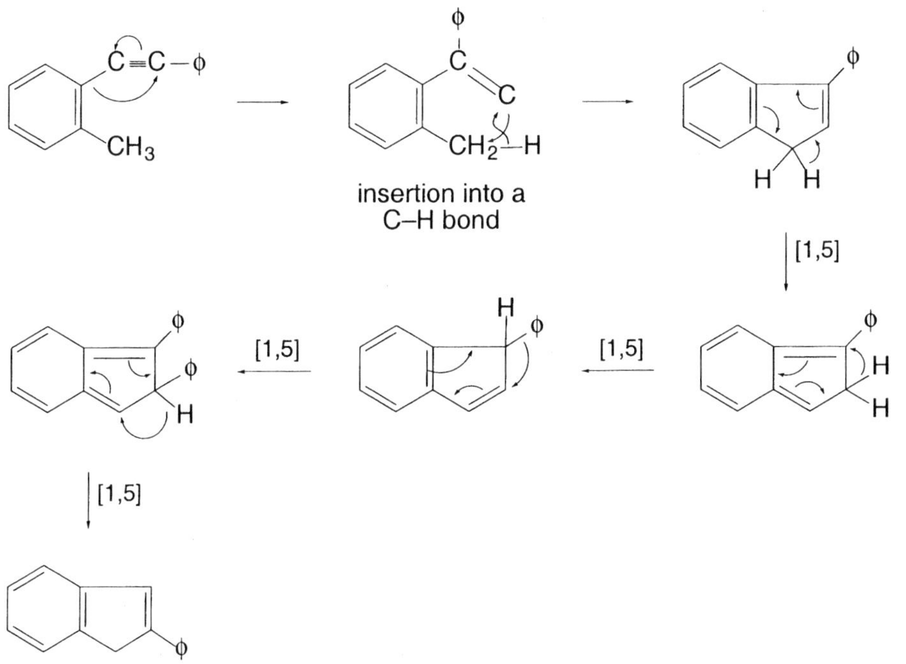

C≡C–ϕ CH_3 → ϕ C C CH_2–H → ϕ H H

insertion into a C–H bond

[1,5]

ϕ H H → [1,5] → H ϕ → [1,5] → ϕ ϕ H

[1,5]

ϕ

3l.

D, CH_3, CH_3, $(H_3C)_2HC$ → $\ddot{C}$, C, $CD(CH_3)_2$, H, C, H_3C, CH_3 → H, $CD(CH_3)_2$, H_3C, CH_3

3m. $(CH_2)_6$, CCl_2 → $(CH_2)_6$, Cl, Cl, CCl_2 → $(CH_2)_6$, Cl, Cl, Cl, Cl

?

$(CH_2)_6$, Cl, Cl, Cl, Cl ← $(CH_2)_6$, Cl, Cl, Cl, Cl ← $(CH_2)_6$, Cl, Cl, Cl, Cl

$(CH_2)_6$, Cl, Cl, Cl, Cl —-HCl→ $(CH_2)_6$, Cl, Cl, Cl

4.

H, :C, C=O, C≡N → C=O, C≡N —[1.5]→ C=O, C≡N

CHAPTER 10

PHOTOCHEMISTRY

PROBLEM SOLUTIONS

1.

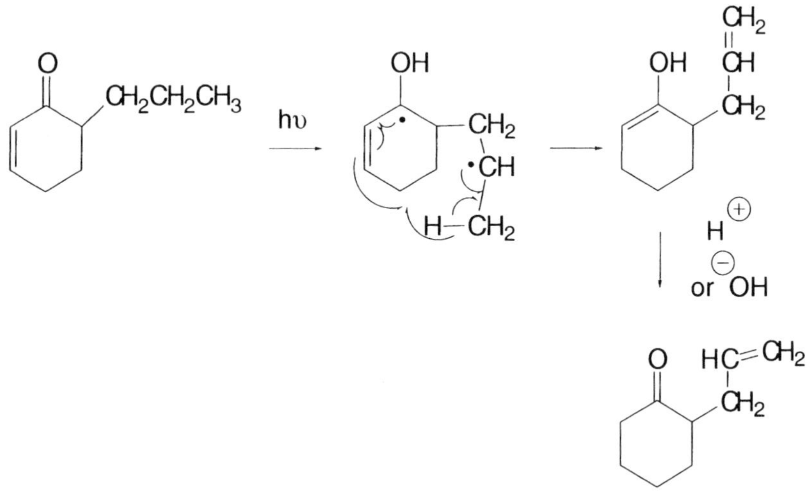

O, $CH_2CH_2CH_3$ — hυ → OH, CH_2, •CH, H–CH_2 → OH, CH_2=CH–CH_2 → ($H^{\oplus}$ or $^{\ominus}OH$) → O, HC=CH_2, CH_2

7

or OH, CH_2, •CH, CH_3 → H_3C, OH → ($H^{\oplus}$ or $^{\ominus}OH$) → H_3C, O

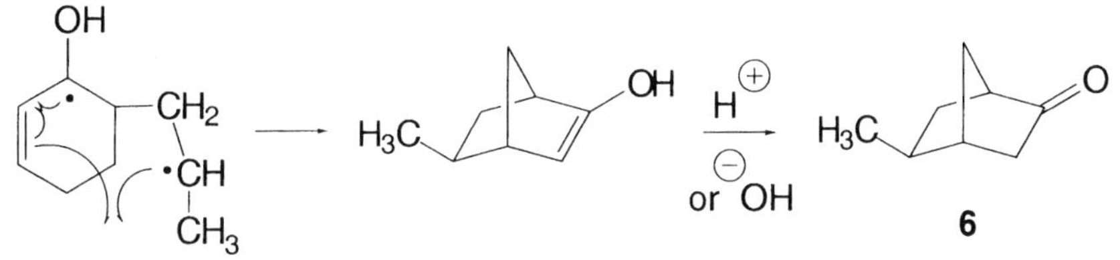

6

2a. O, CH_3, CH_3 ⇌ (hυ) O, CH_3, C–H, CH_3 ⇌ O, CH_3, CH_3

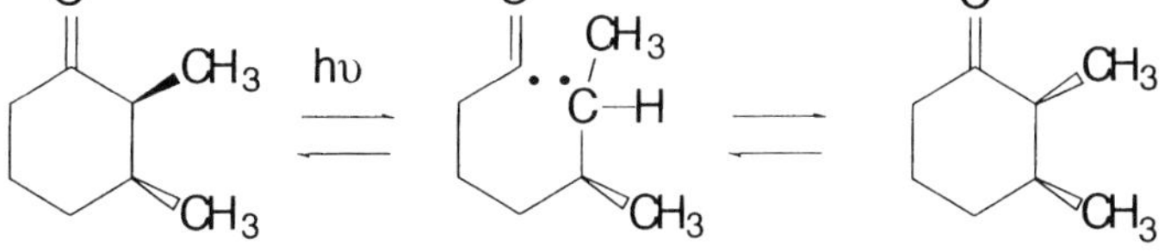

2b.

CH_2 O hυ CH_2 O H_2C C O

CO_2CH_3 O C CH₃ O H O C CH_3 O H

2c.

O hυ O C CH_2 O C CH_2 H_2C O H_2C

2d.

O + OH $H_3C—C—CH_3$ H hυ HO • $C(CH_3)_2$ OH OH $C(CH_3)_2$ OH

$H^{\oplus}$ or $^{\ominus}OH$

O $C(CH_3)_2$ OH

2e.

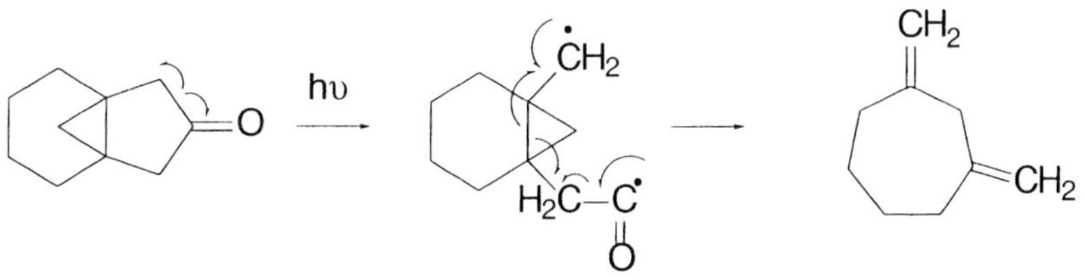
hυ
CH2
H2C—C
O
CH2
CH2

2f.

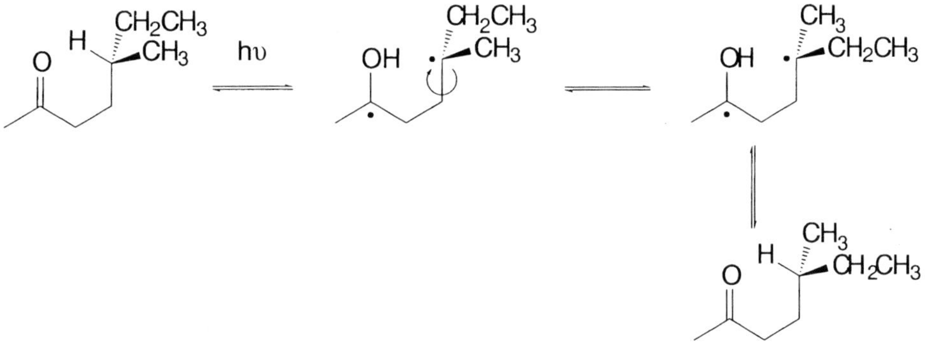
O
H
CH2CH3
CH3
hυ
OH
CH2CH3
CH3
OH
CH3
CH2CH3
O
H
CH3
CH2CH3

2g.

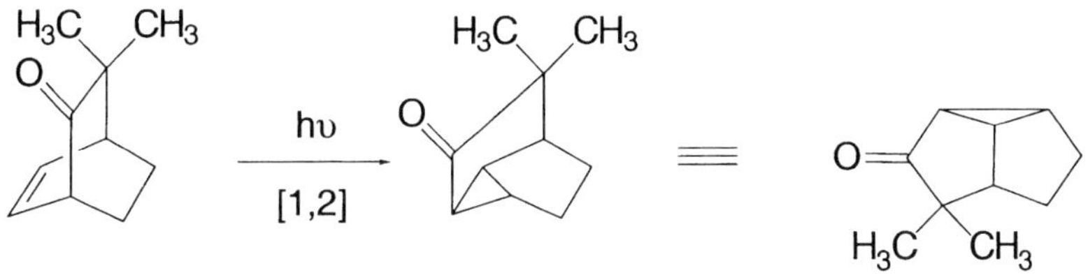
H3C
CH3
O
hυ
[1,2]
H3C
CH3
O
O
H3C
CH3

2h.

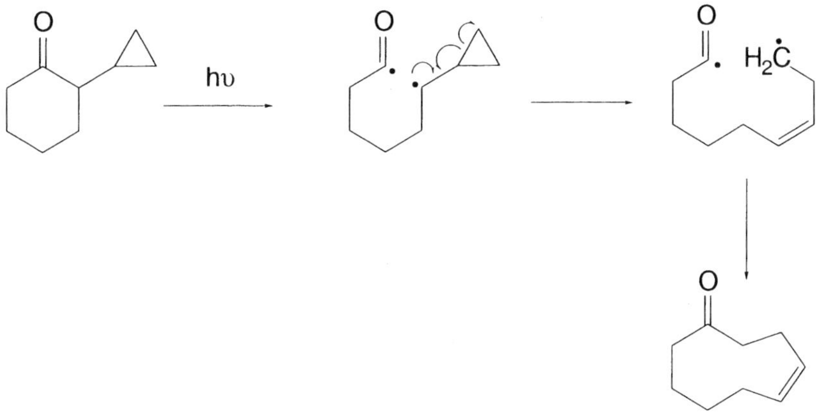
O
hυ
O
O
H2C
O

2i.

O H hυ OH H_3C OH

2j.

hυ OH O OH H C O OH H C O

H O O O

2k.

CH_3 hυ CH_3 CH_3

CH_3

2l.

O C CH $(CH_2)_n$ hυ CH $(CH_2)_n$ C O HC H O C $(CH_2)_n$

The mechanism for the final step in this reaction is uncertain. A direct [1,5] H shift seems less likely than a base or acid catalyzed reaction.

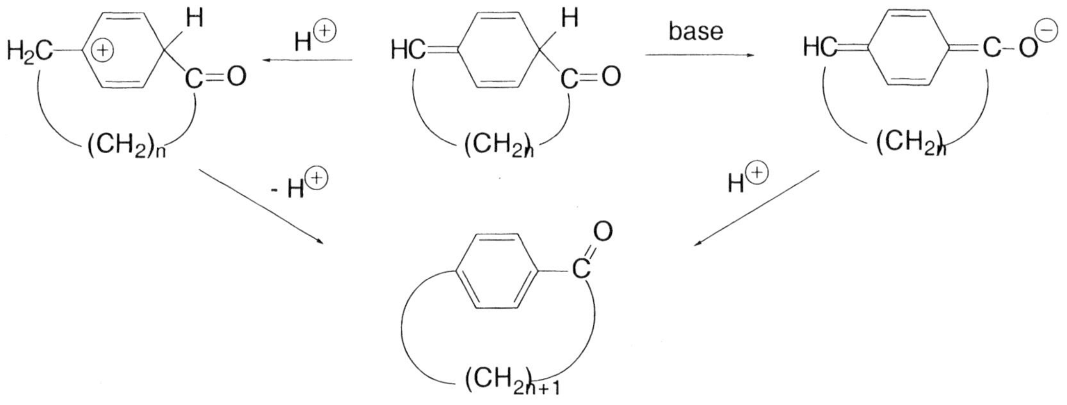

2m.

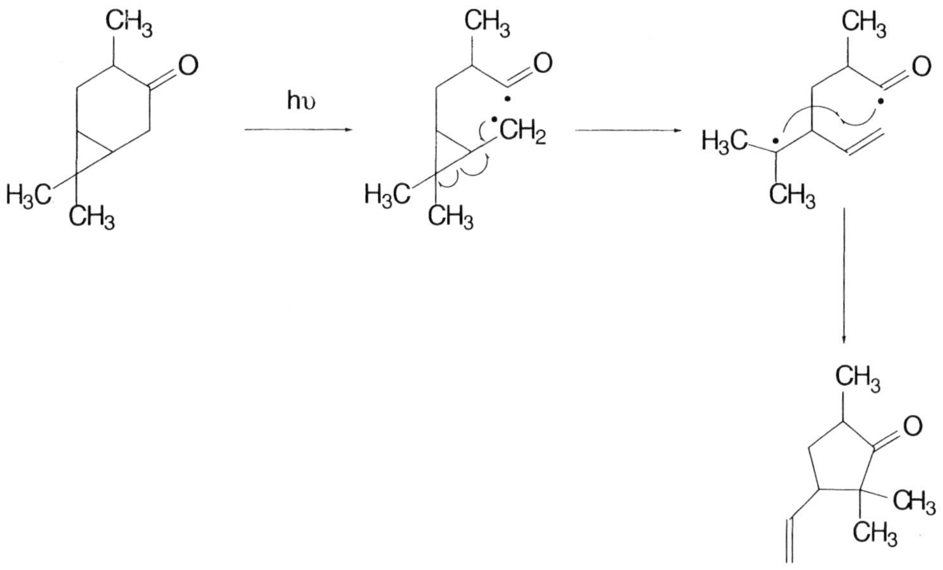

and:

CH_3 O H_3C CH_3 —hυ→ CH_3 O $\dot{C}H_2$ H_3C CH_3 → CH_3 O CH_3 H_3C

↓

H_3C O CH_3 H_3C

CHAPTER 11

SIX-MEMBERED HETEROCYCLIC RINGS

PROBLEM SOLUTIONS

1a. 2-ethylpyridine + CH_3Br ⟶ 1-methyl-2-ethylpyridinium $Br^{\ominus}$

(Structures: pyridine ring with N and C_2H_5; product $\oplus N$–CH_3 with C_2H_5, $Br^{\ominus}$)

1b. quinoline + $C_6H_5\overset{O}{\overset{\|}{C}}OOH$ ⟶ quinoline N-oxide ($N^{\oplus}$–$O^{\ominus}$) + $C_6H_5\overset{O}{\overset{\|}{C}}OH$

1c. 4-chloropyridine (Cl, N) + HNO_3 $\xrightarrow{H_2SO_4}$ 4-chloropyridinium ($\oplus N$–H, Cl) $^{\ominus}NO_3$

1d. 3-chloropyridine (Cl, N) + $NaOCH_3$ ⟶ No Reaction

1e.

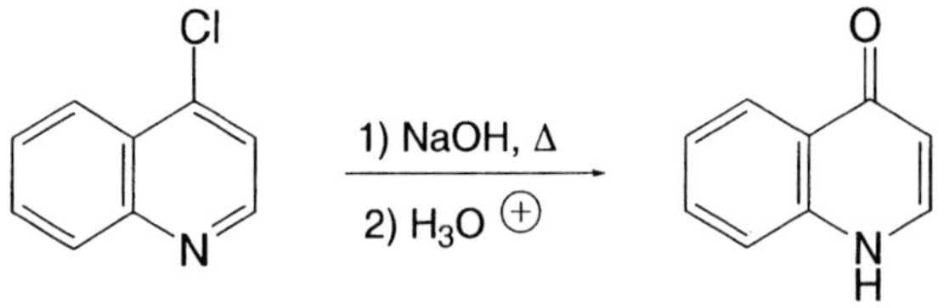

Cl

N

1) NaOH, Δ

2) $H_3O^{\oplus}$

O

N
H

1f.

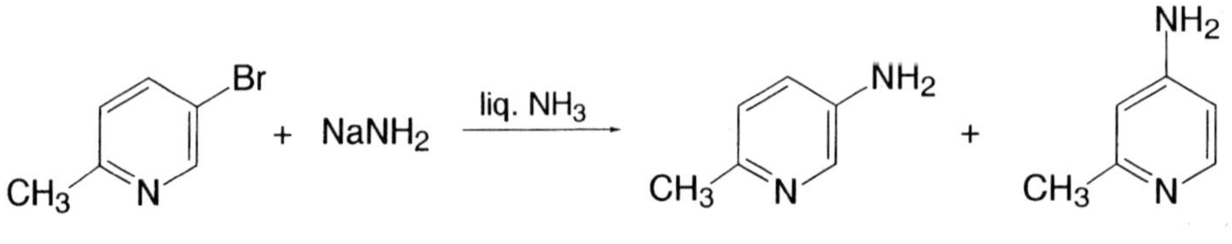

Br

CH_3 N

+ $NaNH_2$

liq. NH_3

NH_2

CH_3 N

+

NH_2

CH_3 N

1g.

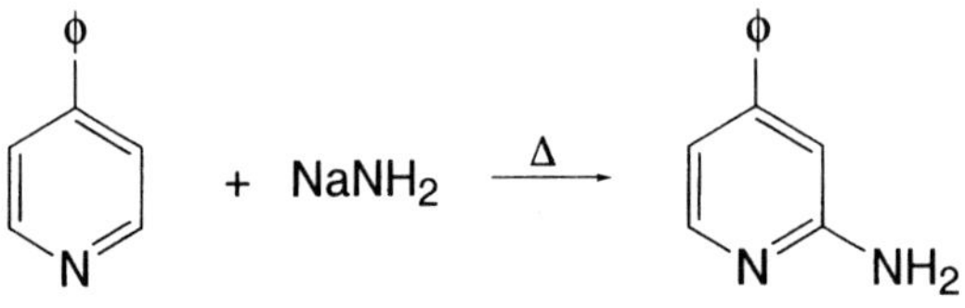

ϕ

N

+ $NaNH_2$

Δ

ϕ

N NH_2

1h.

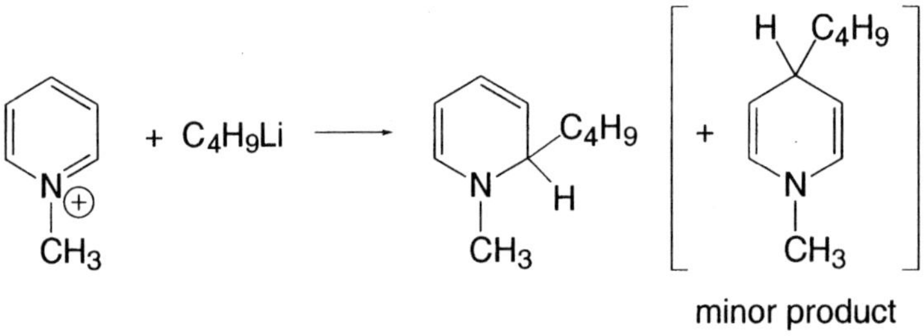

$N^{\oplus}$
CH_3

+ C_4H_9Li

C_4H_9
N H
CH_3

+

H C_4H_9

N
CH_3

minor product

2a.

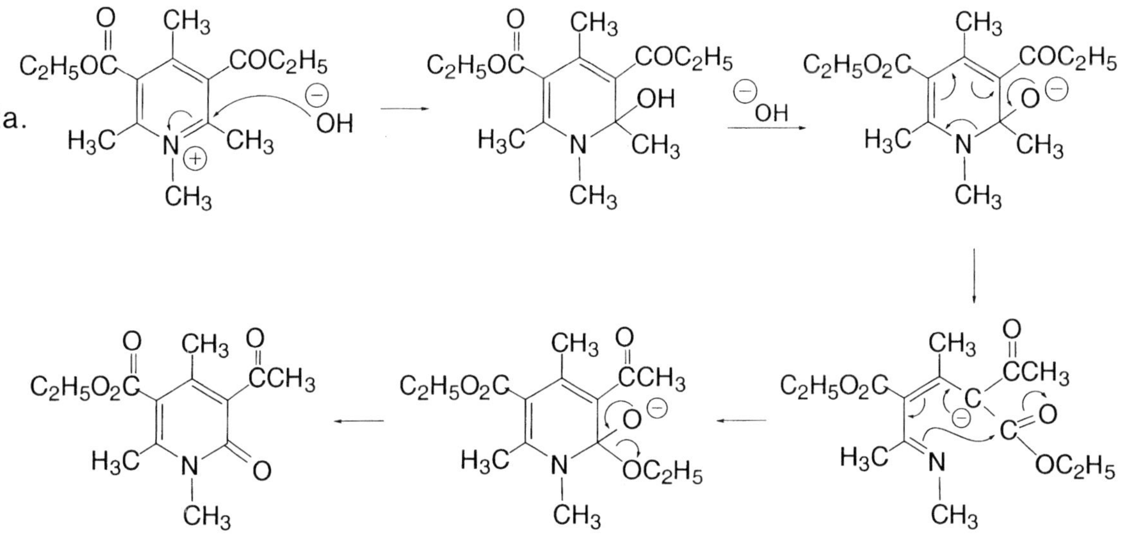

2b.

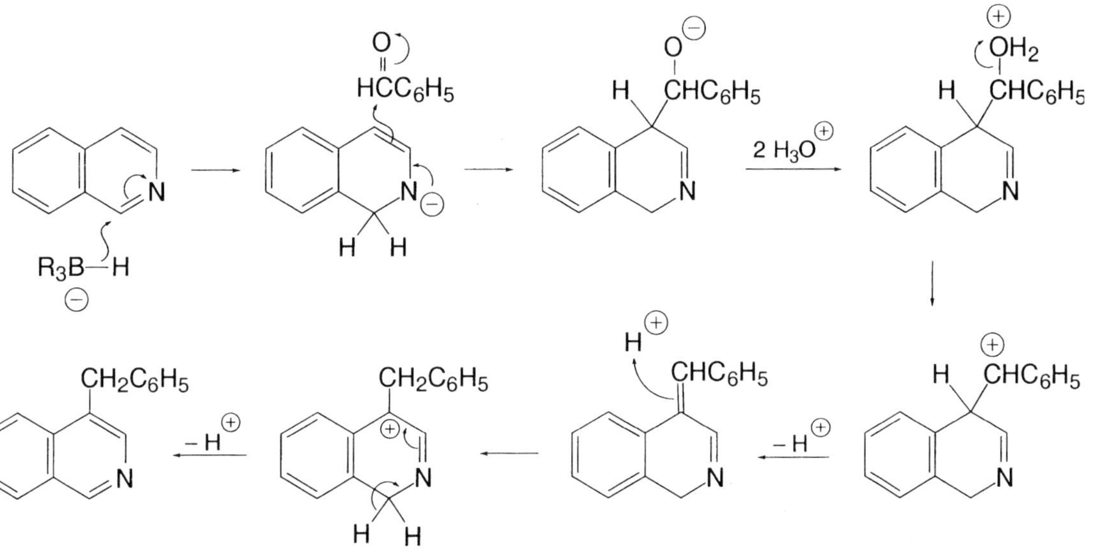

2c.

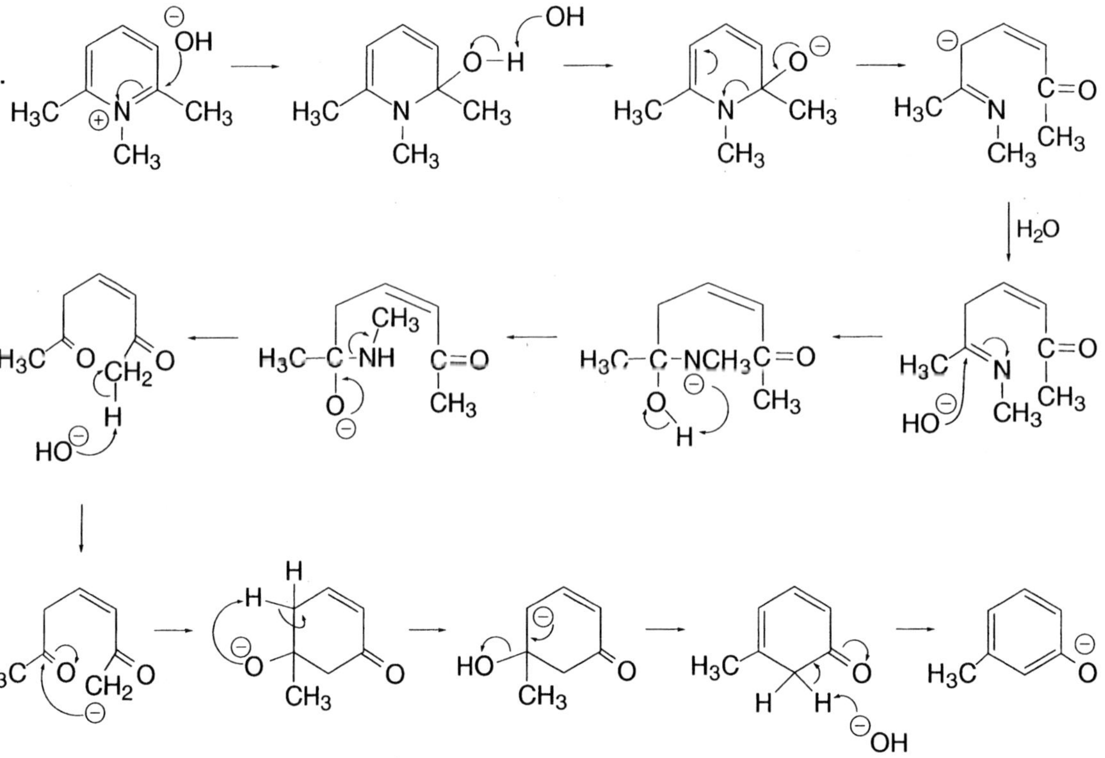

2d.

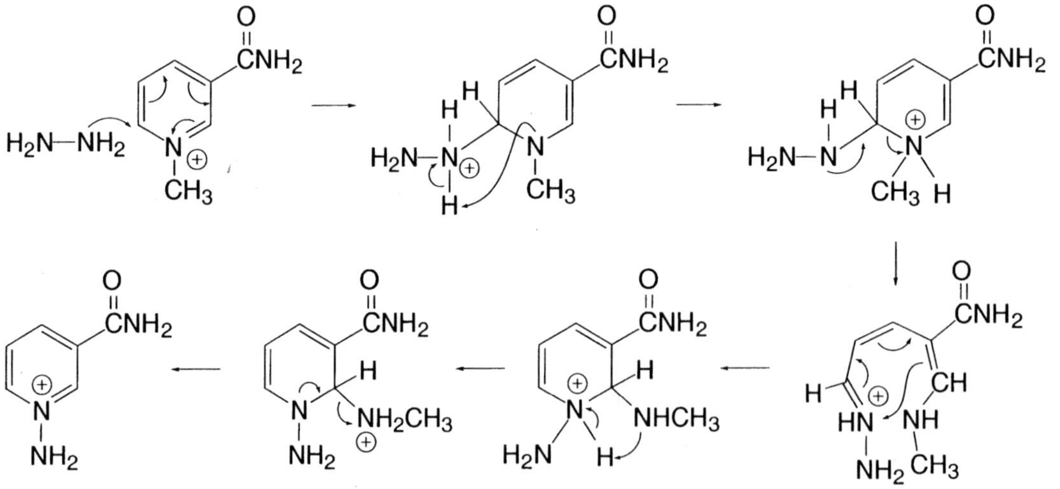

2e. CH_3 CH_3 H C_5H_5 CH_3 CH N CH_3 N H CH_3 NH CH_3

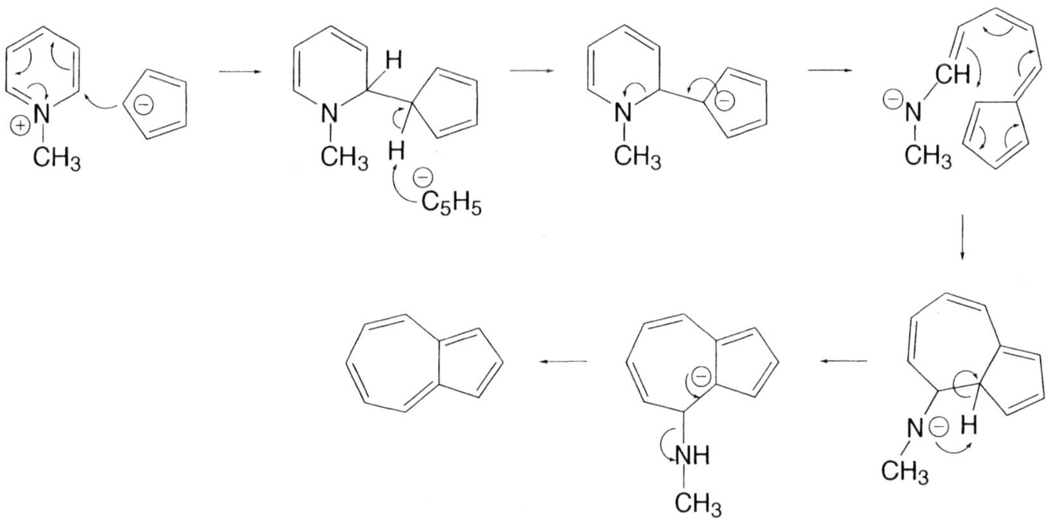

2f. NO_2 N $HONH_2$ H NO_2 N $HO-NH_2$ OH H NO_2 HO N H N H NO_2 HN N H OH H_2O NO_2 H_2N N NO_2 HN N

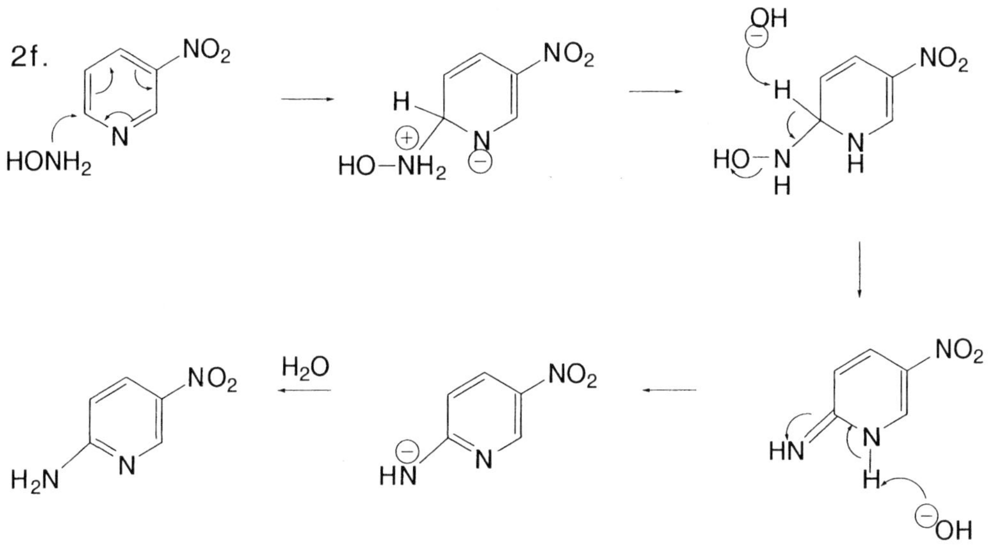

2g.

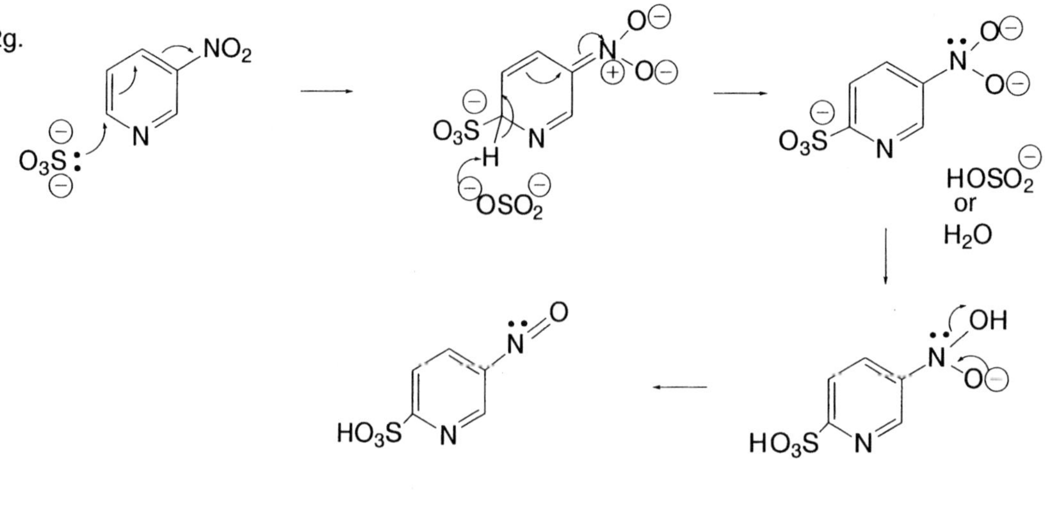
NO2
O3S
N
O3S
H
OSO2
HOSO2
or
H2O
OH
HO3S
O

3. **eq. 52**

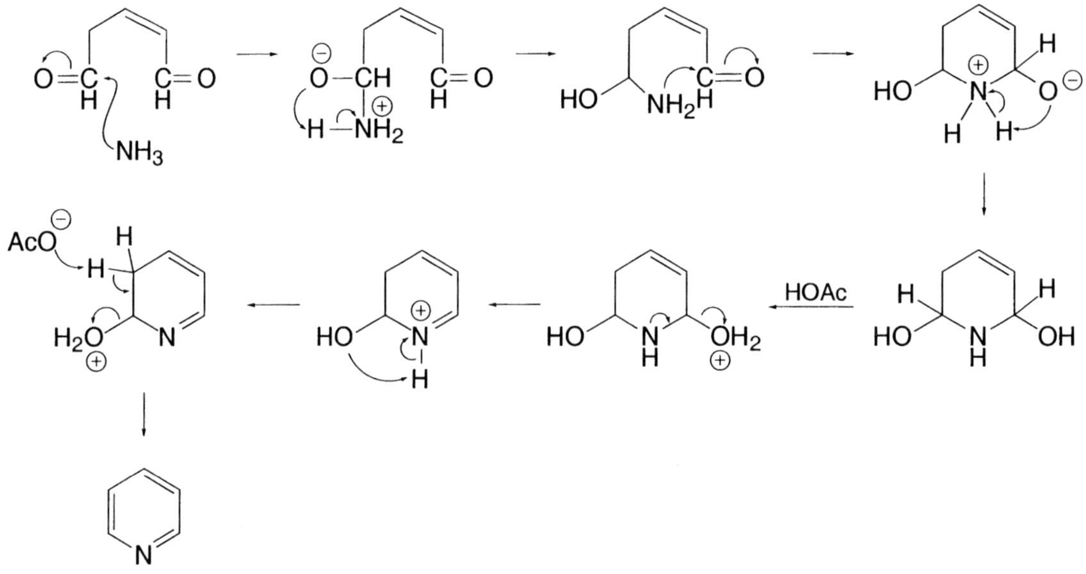
O=C
C=O
H
NH3
O–CH
H–NH2
HO
NH2H
H
AcO
H2O
HOAc
OH
OH2

3. **eq. 53**

C_4H_9 $O=C$ $HC=O$ CH_3 H_2NOH → C_4H_9 $O=C$ $CH-O^{\ominus}$ CH_3 $HN^{\oplus}-H$ OH → C_4H_9 $O=C$ $CH-OH$ CH_3 NH OH

2 steps

C_4H_9 H_3C HO N OH OH ← HOAc ← C_4H_9 $H_2O^{\oplus}$ H_3C N OH OH ← $^{\ominus}OAc$ H H C_4H_9 H_3C $N^{\oplus}$ OH ← C_4H_9 H_3C N OH OH

repeat
last 3 steps

C_4H_9 H_3C N OH → HOAc → H H $^{\ominus}OAc$ C_4H_9 H_3C N $OH_2^{\oplus}$ → C_4H_9 H_3C N

3. **eq. 56**

$CH_3\overset{O}{\overset{\|}{C}}CH_2\overset{O}{\overset{\|}{C}}OC_2H_5$ $\xrightarrow{NH_3}$ $CH_3\overset{O}{\overset{\|}{C}}-\overset{\ominus}{C}H-\overset{O}{\overset{\|}{C}}OC_2H_5$ $O=CCH_3$ H → $CH_3\overset{O}{\overset{\|}{C}}-CH-\overset{O}{\overset{\|}{C}}OC_2H_5$ $^{\ominus}O-CHCH_3$

$\overset{\oplus}{NH_4}$

$CH_3\overset{O}{\overset{\|}{C}}-C-\overset{O}{\overset{\|}{C}}OC_2H_5$ $=CHCH_3$ ← $CH_3\overset{O}{\overset{\|}{C}}-\overset{\ominus}{C}-\overset{O}{\overset{\|}{C}}OC_2H_5$ $HO-CHCH_3$ $\xleftarrow{NH_3}$ $CH_3\overset{O}{\overset{\|}{C}}-CH-\overset{O}{\overset{\|}{C}}OC_2H_5$ $HO-CHCH_3$

and:

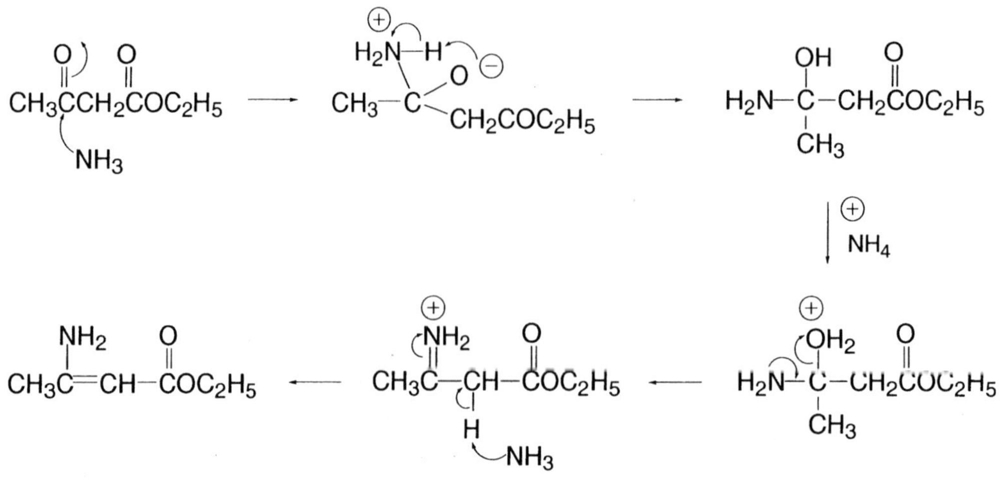

$CH_3CCH_2COC_2H_5$ + NH_3 ⟶ $CH_3-C(O^-)(NH_2^+H)CH_2COC_2H_5$ ⟶ $H_2N-C(OH)(CH_3)-CH_2COC_2H_5$

↓ NH_4^+

$CH_3C(NH_2)=CH-COC_2H_5$ ⟵ $CH_3C(=NH_2^+)-CH-COC_2H_5$ (H, NH_3) ⟵ $H_2N-C(OH_2^+)(CH_3)-CH_2COC_2H_5$

then:

$C_2H_5O_2C-CH=C(CH_3)NH_2$ + $CH_3CH=C(CO_2C_2H_5)C(=O)CH_3$ ⟶ ring-closure intermediate (NH^+, O^-)

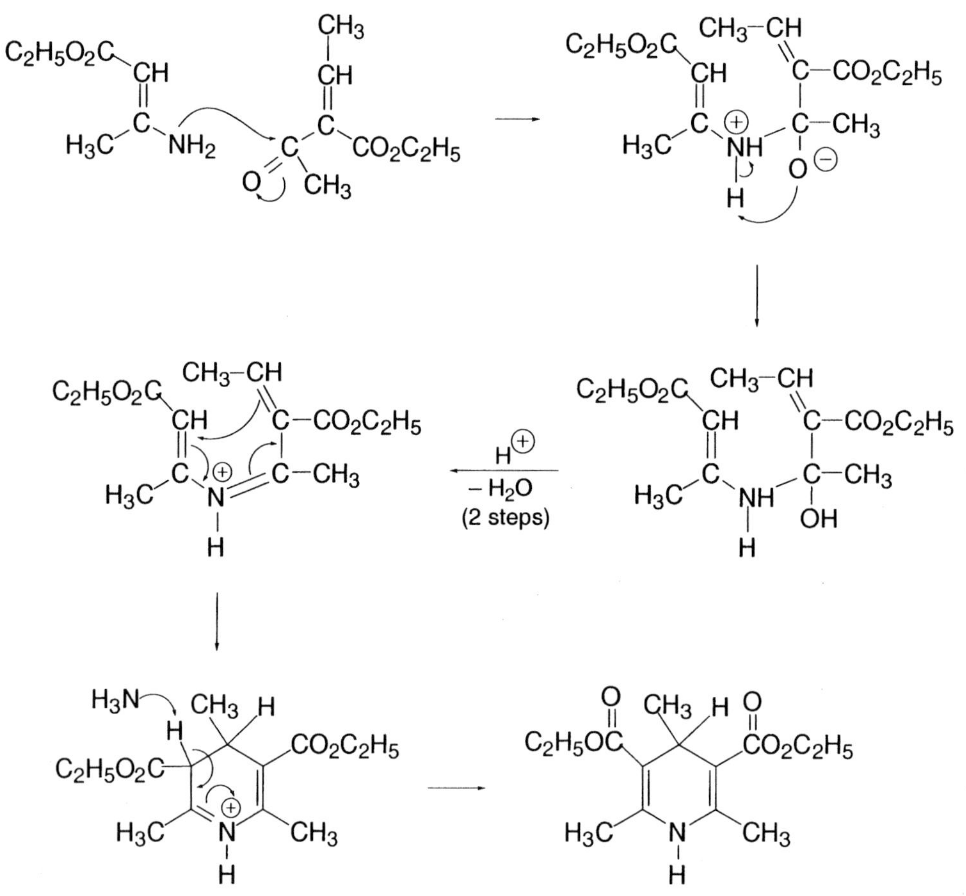

↓

$-H_2O$ (2 steps), H^+

↓

H_3N ... $C_2H_5O_2C$, $CO_2C_2H_5$, H_3C, CH_3, N–H ⟶ $C_2H_5OC(=O)$, CH_3, H, $CO_2C_2H_5$, H_3C, CH_3, N–H

3. **eq. 58**

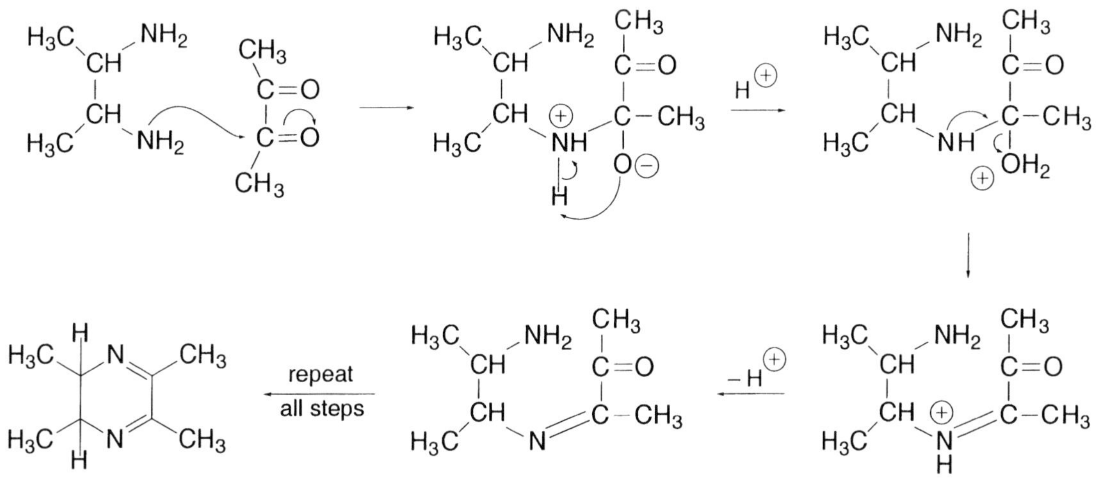

3. **eq. 65**

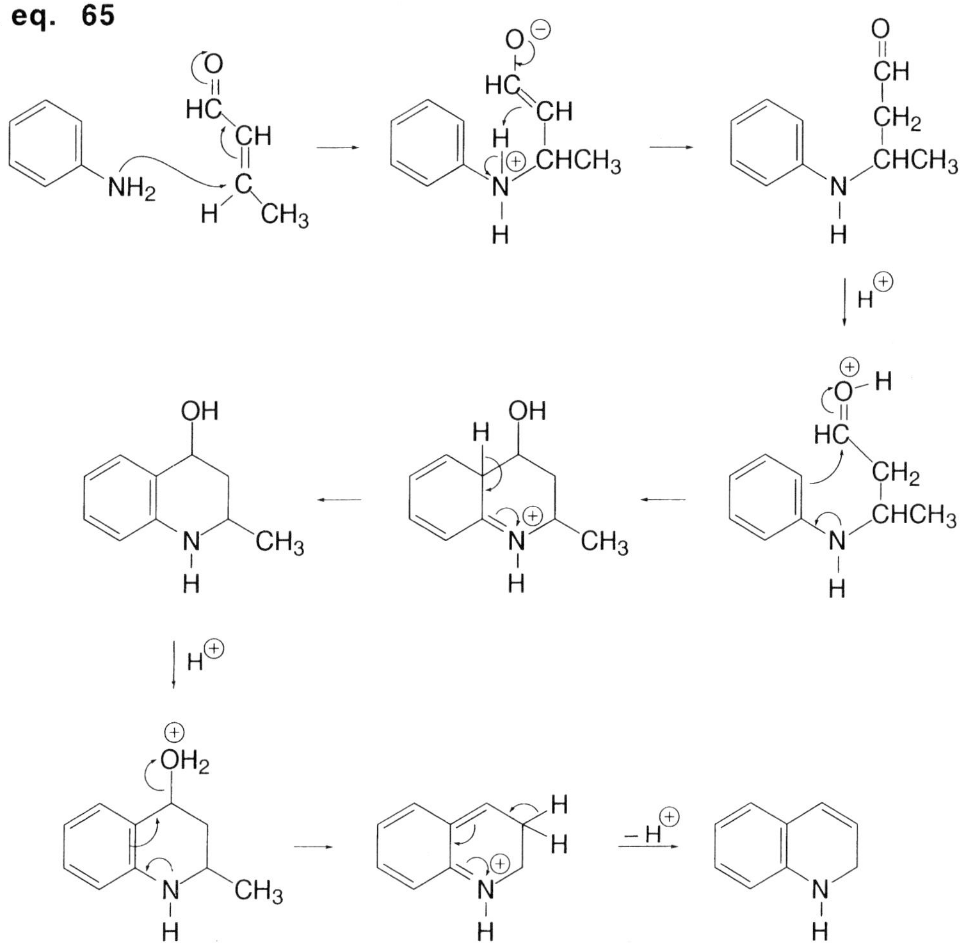

3. eq. 66

$CH_2{-}OH$
$CH{-}OH$ $H{-}OSO_3H$ → $CH_2{-}OH$ / $CH{-}\overset{\oplus}{O}H_2$ / $CH_2{-}OH$ → $^{\ominus}OSO_3H$, H / $CH{-}OH$ / $CH^{\oplus}$ / $CH_2{-}OH$ → $CH{-}OH$ / $\|$ / CH / $CH_2{-}OH$ / $H{-}OSO_3H$
$CH_2{-}OH$

↓

$CH{=}O$ / CH / $\|$ / CH_2 ←($-H^{\oplus}$)— $CH{=}\overset{\oplus}{O}H$ / CH / $\|$ / CH_2 ← $CH{-}OH$ / $\|$ / CH / $CH_2{-}\overset{\oplus}{O}H_2$

Succeeding steps will be similar to those in eq. 65.

3. eq. 67

H_3CO, CH_2, CH_2, NH, C, H_3C, O, O, PCl_2, Cl → H_3CO, CH_2, CH_2, $^{\oplus}NH$, C, O, H_3C, $O{-}PCl_2$ → $H_3CO^{\oplus}$, NH, H, CH_3, $OPCl_2$, O

↓ $-H^{\oplus}$

NH, CH_3 ←($-H^{\oplus}$)— H_3CO, $^{\oplus}NH$, CH_3 ← H_3CO, NH, CH_3, $OPCl_2$, O

3. **eq. 68**

$C_6H_5CH_2CH_2NH_2 + HC(=O)CH_3 \rightarrow C_6H_5CH_2CH_2NH(H)^{\oplus}\text{-}CH(O^{\ominus})CH_3 \rightarrow C_6H_5CH_2CH_2NH\text{-}CH(OH)CH_3$

$\xrightarrow{H_2SO_4} C_6H_5CH_2CH_2NH\text{-}CH(OH_2^{\oplus})CH_3 \rightarrow C_6H_5CH_2CH_2NH^{\oplus}{=}CHCH_3 \rightarrow$ (cyclized cation, H, H, CH_3, NH) $\xrightarrow{-H^{\oplus}}$ 1-methyl-1,2,3,4-tetrahydroisoquinoline (NH, CH_3)

CHAPTER 12

FIVE-MEMBERED HETEROCYCLIC RINGS

PROBLEM SOLUTIONS

1a. pyrrole + CH_3Cl ⟶ No Reaction

1b. 4-methylimidazole + C_2H_5I ⟶ 1-ethyl-5-methylimidazolium iodide ($N-C_2H_5$, CH_3, N–H, $I^{\ominus}$)

1c. benzothiophene + HNO_3 $\xrightarrow{Ac_2O}$ 3-nitrobenzothiophene (NO_2)

1d. pyrrole $\xrightarrow[2)\ Ac_2O]{1)\ NaNH_2}$ N-acetylpyrrole + HOAc

1e. 1-methyl-2-phenylpyrrole (ϕ, CH_3) + $CH_3\overset{O}{\overset{\|}{C}}H$ $\xrightarrow{HCl}$ CH_3CH-bis(1-methyl-5-phenylpyrrol-2-yl) (ϕ, CH_3, CH, CH_3, CH_3, ϕ)

1f. (furan) + (maleic anhydride) ⟶ (H, H; O; O; O; O)

1g. (indole, N–H) + $CHCl_3$ + $\overset{\oplus}{K}$ $\overset{\ominus}{O}$–$C(CH_3)_3$ ⟶ (3-chloroquinoline: Cl, N) + $HOC(CH_3)_3$

1h. ϕ–$\overset{O}{\overset{\|}{C}}$–$CH_2NH_2$ + $CH_3O\overset{O}{\overset{\|}{C}}CH_2\overset{O}{\overset{\|}{C}}\phi$ $\xrightarrow{HOAc}$ (pyrrole: ϕ, $\overset{O}{\overset{\|}{C}}OCH_3$, ϕ, N–H)

1i. (phenyl)–$\overset{H}{\overset{|}{N}}$–N=C($\phi$)($CH_3$) $\xrightarrow{H_2SO_4}$ (indole: N–H, ϕ)

1j. $CH_3O\overset{O}{\overset{\|}{C}}$–CH=$\overset{\oplus}{N}$=$\overset{\ominus}{N}$ + $CH_3O\overset{O}{\overset{\|}{C}}$–C≡C–$\overset{O}{\overset{\|}{C}}OCH_3$ ⟶ (pyrazole: $CH_3O\overset{O}{\overset{\|}{C}}$, $\overset{O}{\overset{\|}{C}}OCH_3$, $CH_3O\overset{O}{\overset{\|}{C}}$, N, N–H)

2. Pyrazole should be (and in fact is) a much stronger base than pyrrole, because protonation of the doubly bonded nitrogen in pyrazole yields a cation with appreciable resonance stabilization.

NH ⊕ N H ⟷ ⊕ NH N H

In contrast, pyrazole should be (and is) a weaker base than imidazole. The resonance stabilizations of the two cations should be similar, but the basicity of pyrazole is decreased by the inductive effect of the nitrogen-nitrogen bond.

3. **eq. 55**

CH_2-CH_2 C C_6H_5 O O C C_6H_5 $\xrightarrow{H^{\oplus}}$ ⊕ HO=C CH_2-CH_2 C C_6H_5 O C_6H_5 ⟶ HO H H C_6H_5 O ⊕ C_6H_5

$-H^{\oplus}$

H H C_6H_5 O ⊕ C_6H_5 ⟵ ⊕ H_2O C_6H_5 O C_6H_5 $\xleftarrow{H^{\oplus}}$ HO C_6H_5 O C_6H_5

$H^{\oplus}$

C_6H_5 O C_6H_5

3. **eq. 57**

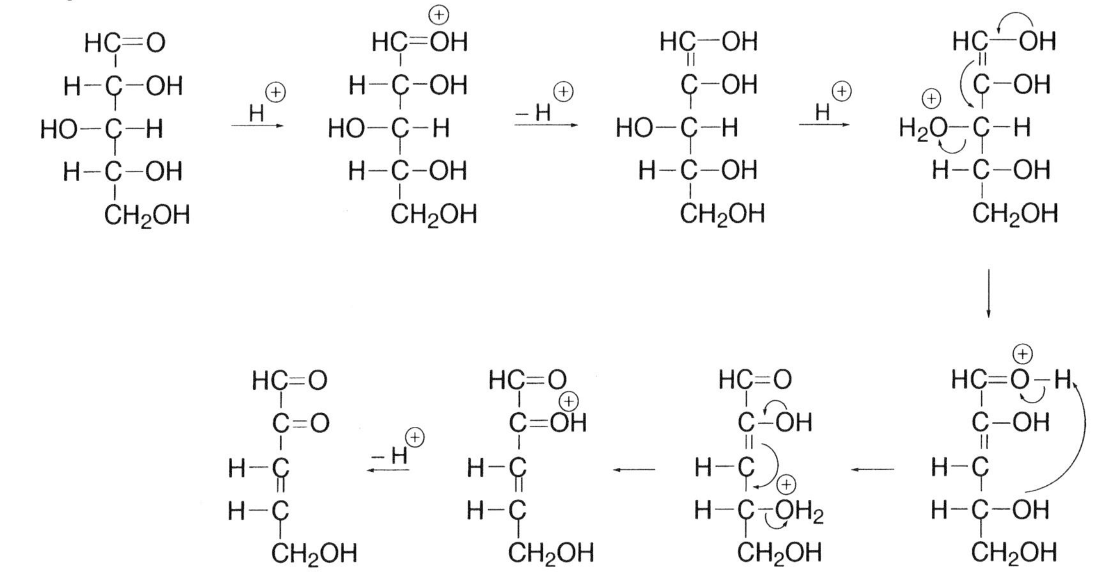

HC=O / H–C–OH / HO–C–H / H–C–OH / CH₂OH →(H⊕) HC=OH⊕ / H–C–OH / HO–C–H / H–C–OH / CH₂OH →(–H⊕) HC–OH / C–OH / HO–C–H / H–C–OH / CH₂OH →(H⊕) HC–OH / C–OH / ⊕H₂O–C–H / H–C–OH / CH₂OH

↓

HC=O⊕–H / C–OH / H–C / H–C–OH / CH₂OH → HC=O / C–OH / H–C / H–C–OH₂⊕ / CH₂OH → HC=O / C=OH⊕ / H–C / H–C / CH₂OH →(–H⊕) HC=O / C=O / H–C / H–C / CH₂OH

(While dehydration may also occur by a simple E1 mechanism, loss of water from positions β to carbonyl groups can take place under unusually mild conditions because the carbonyls can first be converted to their enol tautomers, as shown above.)

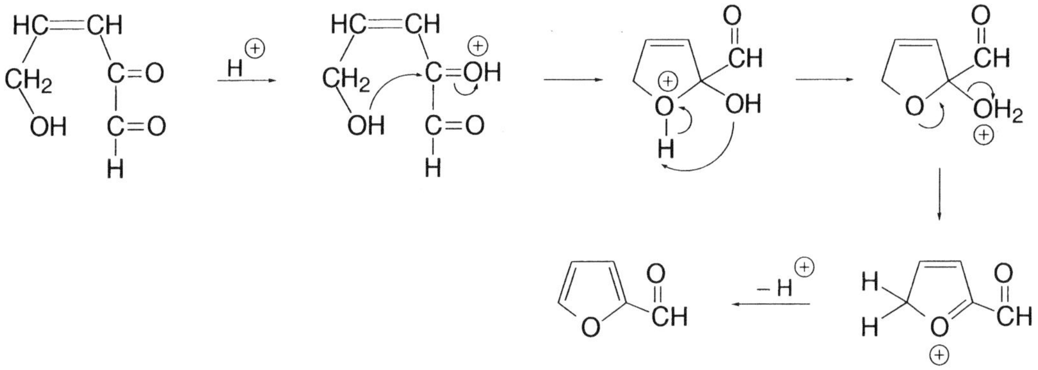

HC=CH / CH₂ C=O / OH C=O / H →(H⊕) HC=CH / CH₂ C=OH⊕ / OH C=O / H → (cyclic oxonium, CH=O, OH) → (cyclic ether, CH=O, OH₂⊕)

↓

(furan-2-carbaldehyde) ←(–H⊕) (H, H, O⊕ ring, CH=O)

3. **eq. 60**

$$\phi\text{-C(=O)-CH}_2\text{-NH}_2 + \text{CH}_3\text{-C(=O)-CH}_2\text{-COC}_2\text{H}_5\text{(=O)} \longrightarrow \phi\text{-C(=O)-CH}_2\text{-N}^{\oplus}\text{H}_2\text{-C(CH}_3\text{)(O}^{\ominus}\text{)-CH}_2\text{-COC}_2\text{H}_5\text{(=O)} \longrightarrow \phi\text{-C(=O)-CH}_2\text{-NH-C(CH}_3\text{)(OH)-CH}_2\text{-COC}_2\text{H}_5\text{(=O)}$$

$$\xrightarrow{\text{HOAc}} \phi\text{-C(=O)-CH}_2\text{-NH-C(CH}_3\text{)(}\overset{\oplus}{\text{O}}\text{H}_2\text{)-CH}_2\text{-COC}_2\text{H}_5\text{(=O)} \longrightarrow \phi\text{-C(=O)-CH}_2\text{-}\overset{\oplus}{\text{N}}\text{H=C(CH}_3\text{)-CH}_2\text{-COC}_2\text{H}_5\text{(=O)} \xrightarrow{-\text{H}^{\oplus}} \phi\text{-C(=O)-CH}_2\text{-NH-C(CH}_3\text{)=CH-CO}_2\text{C}_2\text{H}_5$$

$$\xrightarrow{\text{H}^{\oplus}} \text{H}\overset{\oplus}{\text{O}}\text{=C(}\phi\text{)-CH}_2\text{-NH-C(CH}_3\text{)=CH-CO}_2\text{C}_2\text{H}_5 \longrightarrow \text{(cyclic iminium: HO, }\phi\text{, H, CO}_2\text{C}_2\text{H}_5\text{, CH}_3\text{, }\overset{\oplus}{\text{N}}\text{-H)} \xrightarrow{-\text{H}^{\oplus}} \text{(HO, }\phi\text{, CO}_2\text{C}_2\text{H}_5\text{, CH}_3\text{, N-H)}$$

$$\xrightarrow{\text{HOAc}} \text{(H}_2\overset{\oplus}{\text{O}}\text{, }\phi\text{, CO}_2\text{C}_2\text{H}_5\text{, CH}_3\text{, N-H)} \longrightarrow \text{(H, H, }\phi\text{, CO}_2\text{C}_2\text{H}_5\text{, CH}_3\text{, }\overset{\oplus}{\text{N}}\text{-H)} \xrightarrow{-\text{H}^{\oplus}} \text{(pyrrole: }\phi\text{, CO}_2\text{C}_2\text{H}_5\text{, CH}_3\text{, N-H)}$$

3. **eq. 61**

$$C_6H_5CH_2-\overset{O}{\overset{\|}{C}}CH_3 \xrightarrow{^{\ominus}OC_2H_5} C_6H_5\overset{\ominus}{C}H-\overset{O}{\overset{\|}{C}}CH_3 \xrightarrow{O=N-OC_2H_5} C_6H_5CH(N(O^{\ominus})OC_2H_5)-\overset{O}{\overset{\|}{C}}CH_3 \longrightarrow C_6H_5CH(H)(N{=}O)-\overset{O}{\overset{\|}{C}}CH_3 \xrightarrow{^{\ominus}OC_2H_5} C_6H_5CH-C(=N-O^{\ominus})-\overset{O}{\overset{\|}{C}}-CH_3 \xrightarrow{C_2H_5OH} C_6H_5CH-C(=N-OH)-\overset{O}{\overset{\|}{C}}-CH_3$$

One possible mechanism for the reduction step begins:

$$C_6H_5C(=N-OH)-\overset{O}{\overset{\|}{C}}CH_3 \xrightarrow{HCl} C_6H_5C(=\overset{\oplus}{N}H-OH)-\overset{O}{\overset{\|}{C}}CH_3 \xrightarrow{Zn:} C_6H_5C(Zn^{\oplus})(NH-OH)-\overset{O}{\overset{\|}{C}}CH_3 \xrightarrow{Cl-H} C_6H_5CH(NH-OH)-\overset{O}{\overset{\|}{C}}CH_3$$

or:

$$C_6H_5C(=\overset{\oplus}{N}H-OH)-\overset{O}{\overset{\|}{C}}CH_3 + Zn: \longrightarrow C_6H_5\dot{C}(\ddot{N}H-OH)-\overset{O}{\overset{\|}{C}}CH_3 + \overset{\oplus}{\dot{Z}n} \longrightarrow C_6H_5C(\oplus Zn)(NH-OH)-\overset{O}{\overset{\|}{C}}CH_3 \xrightarrow{HCl} C_6H_5CH(NH-OH)-\overset{O}{\overset{\|}{C}}CH_3$$

then:

$$C_6H_5CH(NH{-}OH)\overset{\displaystyle O}{\overset{\|}{C}}CH_3 \xrightarrow{H^{\oplus}} C_6H_5CH(NH{-}OH)\overset{\displaystyle \overset{\oplus}{O}H}{\overset{\|}{C}}CH_3 \xrightarrow{-H^{\oplus}} C_6H_5C(NH{-}OH){=}C(OH)CH_3 \xrightarrow{HCl} C_6H_5C(NH{-}\overset{\oplus}{O}H_2){=}C(OH)CH_3$$

$$\longrightarrow C_6H_5\overset{\displaystyle NH}{\overset{\|}{C}}{-}\overset{\displaystyle \overset{\oplus}{O}H}{\overset{\|}{C}}CH_3 \longrightarrow C_6H_5\overset{\displaystyle NH}{\overset{\|}{C}}{-}\overset{\displaystyle O}{\overset{\|}{C}}CH_3 \xrightarrow[\text{as in eqs. above}]{\text{Zn, HCl}} C_6H_5\overset{\displaystyle NH_2}{\overset{|}{C}}{-}H{-}\overset{\displaystyle O}{\overset{\|}{C}}CH_3$$

3. **eq. 62**

Several mechanisms can account for this reaction. One possibility is:

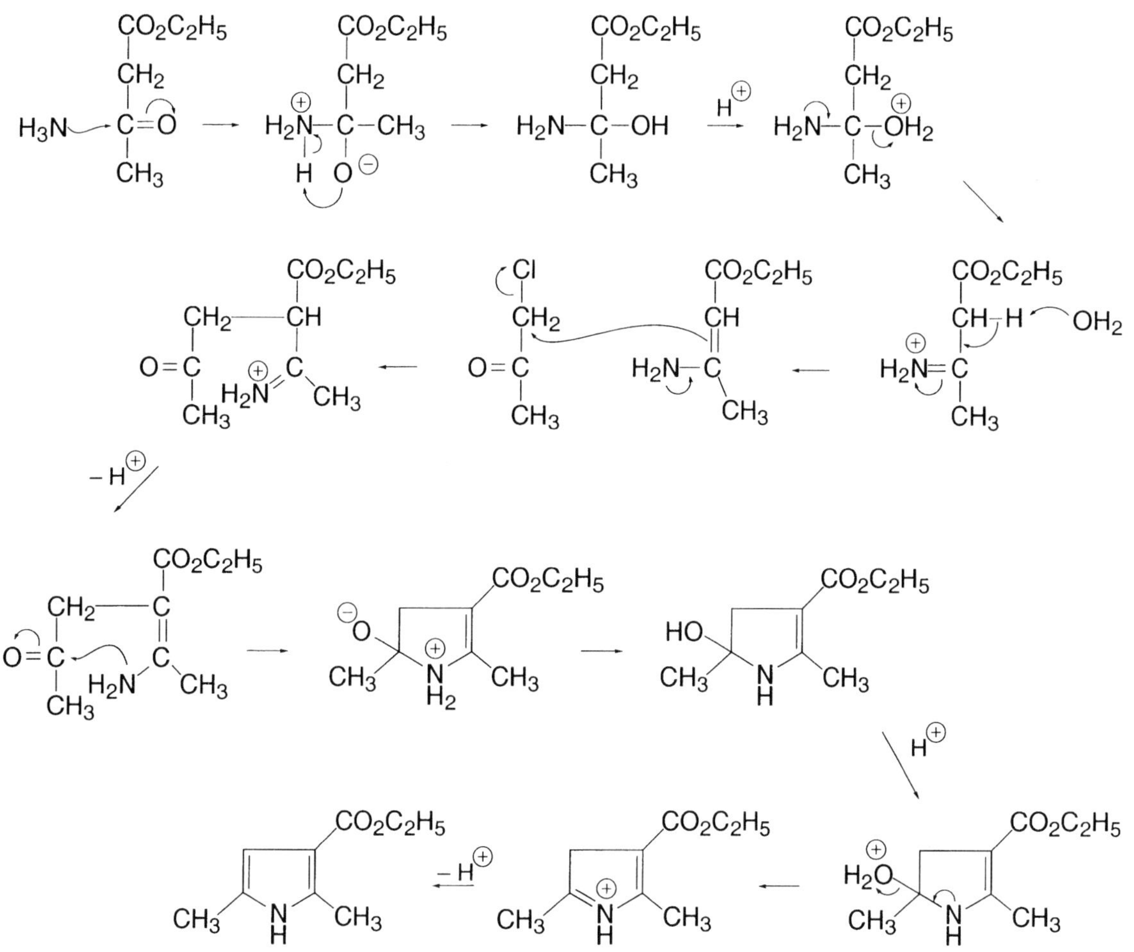

Alternatively, the ring might be formed by an electrocyclic reaction.

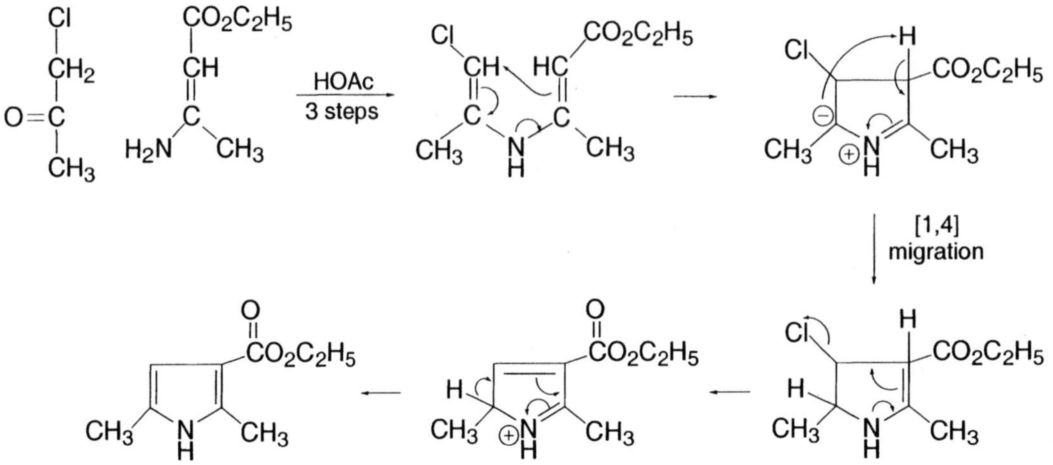

HOAc
3 steps

[1,4]
migration

4a. 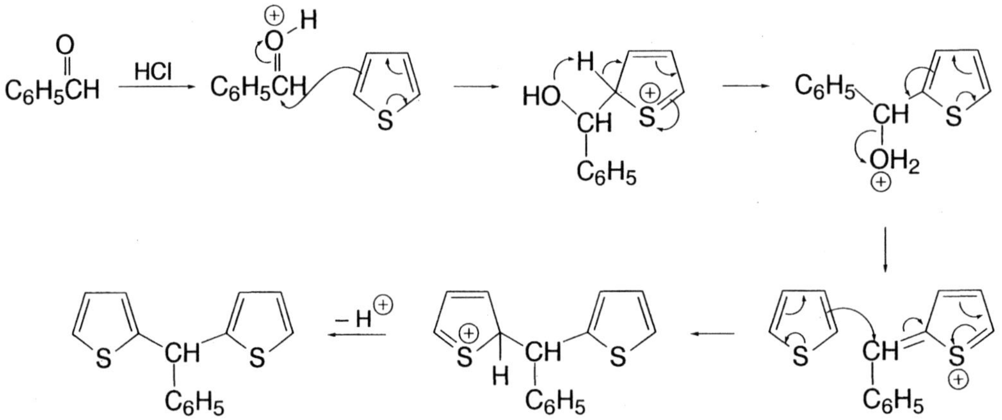 C_6H_5CHO HCl

$-H^{\oplus}$

4b.

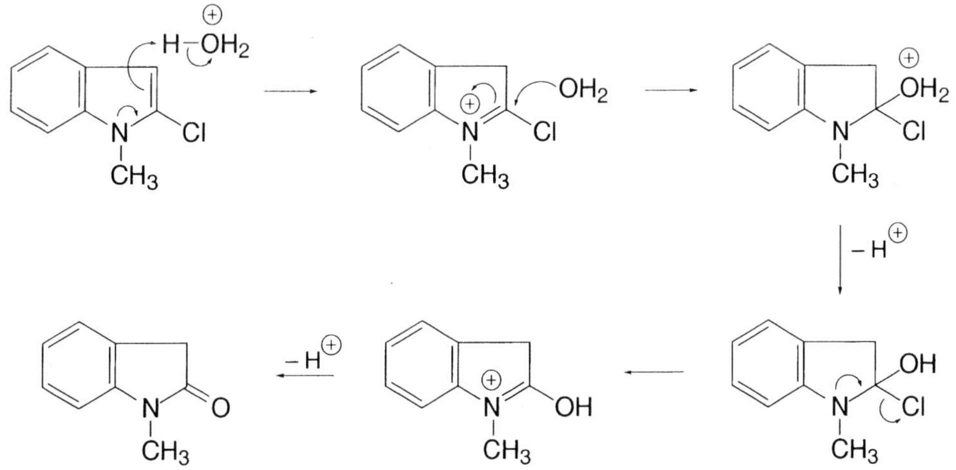

$H-OH_2^{\oplus}$; N, CH_3, Cl → OH_2 ; $\overset{\oplus}{N}$, CH_3, Cl → $OH_2^{\oplus}$; N, Cl, CH_3

$-H^{\oplus}$

OH ; N, Cl, CH_3 → $\overset{\oplus}{N}$, OH, CH_3 → $-H^{\oplus}$ → N, O, CH_3

4c.

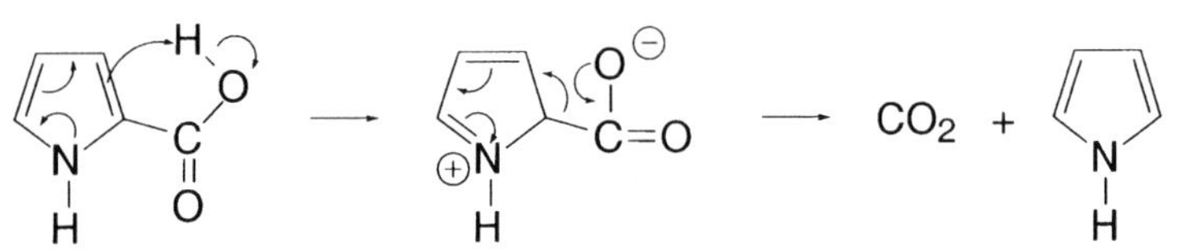

H, O, C, O ; N, H → $O^{\ominus}$, C=O ; $\overset{\oplus}{N}$, H → CO_2 + N, H

4d.

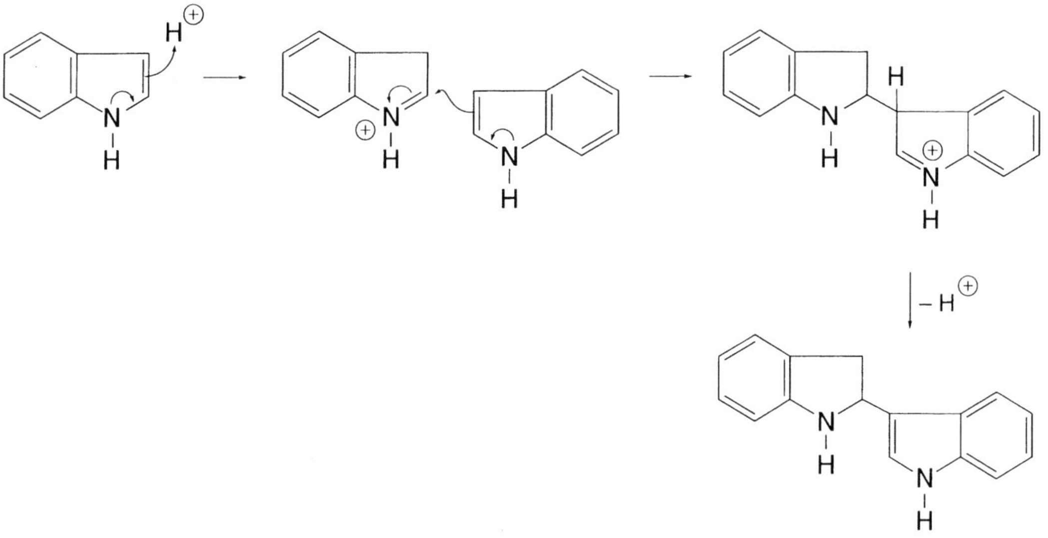

$H^{\oplus}$; N, H → $\overset{\oplus}{N}$, H ; N, H → H ; N, H ; $\overset{\oplus}{N}$, H

$-H^{\oplus}$

N, H ; N, H

4e.

CH_2–OTs
N
CH_3
→
OH_2
N
H
→
OH_2
N
H
$-H^{\oplus}$ →
OH
N
H

4f.

CH_3
ϕ
N
CH_3
$H^{\oplus}$ →
CH_3
ϕ
N
H
CH_3
→
CH_3
CH_3
N
H
ϕ
↓
CH_3
CH_3
N
H
ϕ
$-H^{\oplus}$ ←
CH_3
CH_3
N
ϕ

4g.

CH_3
H
N–N=CHCH$_3$
CH_3
1) $H^{\oplus}$
2) $-H^{\oplus}$
→
CH_3 H_2C=CH
NH
N
H
CH_3
→
CH_3
H
CH_2–C=NH
=NH
CH_3
↓ $H^{\oplus}$
CH_3
H
CH_2–C=NH
NH_2
CH_3
←
H
N
‖
CH
CH_3 H
CH_2
NH_2
CH_3
←
CH_3
CH_2
CH
‖
NH_2 NH_2
CH_3

CH_3
H
N
NH_2
CH_3 H H
→
CH_3
H
N
NH_3
CH_3 H
→
CH_3
OH_2
H
H
N
CH_3 H
→
CH_3
N
CH_3 H

CHAPTER 13

ORGANOPHOSPHORUS AND ORGANOSULFUR CHEMISTRY

PROBLEM SOLUTIONS

1a. $(CH_3)_2P(=O)S^{-}$ + $ClCH_2-C_6H_4-C(=O)Cl$ ⟶ $(CH_3)_2P(=O)-S-CH_2-C_6H_4-C(=O)-O-P(=S)(CH_3)_2$

The soft sulfur atom attacks the soft alkyl carbon, while the hard oxygen atom attacks the hard carbonyl carbon.

1b. $2(CH_3O)_3P$ + $ClCH_2-C_6H_4-C(=O)CH_2Cl$ ⟶ $(CH_3O)_2P(=O)-CH_2-C_6H_4-C(=CH_2)-O-P(=O)(OCH_3)_2$

1c. $CH_3S(=O)-CH_2-C_6H_5$ + CH_3OTs ⟶ $CH_2{=}O$ + $CH_3S-CH_2-C_6H_5$

1d. $2CH_3S(=O)CH_3$ + $ClC(=O)-C_6H_4-C(=O)CH_2Cl$ ⟶ $HOC(=O)-C_6H_4-C(=O)-C(=O)H$ + CH_3SCH_2Cl

1e. $HC(=O)CH_2CH_2CH_2Br$ + $(C_6H_5)_3P$ ⟶ $HC(=O)(CH_2)_3\overset{+}{P}(C_6H_5)_3$ Br^-

NaH ↓

$(C_6H_5)_3P{=}O$ + □ ⟵ [$HC(=O)CH_2CH_2C{=}P(C_6H_5)_3$]

1f. $(C_6H_5)_3\overset{+}{P}{-}CH{=}CH_2$ + $CH_3C(=O){-}\overset{-}{C}H_2{-}C(=O)OCH_3$ ⟶ $(C_6H_5)_3P{=}CH{-}CH_2{-}CH(C(=O)CH_3){-}C(=O)OCH_3$

↓

H_3C, $COCH_3$ (cyclobutene) + $(C_6H_5)_3P{=}O$

2a. $(C_6H_5)_3P{:}$ Br–Br ⟶ $(C_6H_5)_3\overset{+}{P}{-}Br$ Br^- + $H{-}O{-}CH_2CH_3$ ⟶ $(C_6H_5)_3\overset{+}{P}{-}\overset{+}{O}(H){-}CH_2CH_3$ Br^-

↓

$BrCH_2CH_3$ + $(C_6H_5)_3P{=}O$ + HBr

2b. Analogy with the Perkow reaction suggests the mechanism below.

C_6H_5-CH—CH—COC$_2$H$_5$ (epoxide, C=O) + $(C_6H_5)_3P$: → (150°) C_6H_5-CH—CH—C(O$^-$)(OC$_2$H$_5$)—$P^+(C_6H_5)_3$

→ C_6H_5-CH(O$^-$)—CH=C(O—$P^+(C_6H_5)_3$)—OC$_2$H$_5$

→ C_6H_5-CH(O—$P^+(C_6H_5)_3$)—CH^-—C(=O)—OC$_2$H$_5$

→ C_6H_5-CH=CH—C(=O)—OC$_2$H$_5$ + $(C_6H_5)_3P=O$

2c. Tetrahydrothiopyran-4-one S-oxide + $Si(CH_3)_3Cl$ → S$^+$-OSi(CH$_3$)$_3$ intermediate + Cl^- → sulfonium ion + $^-OSi(CH_3)_3$ → 2,3-dihydro-4H-thiopyran-4-one + $HOSi(CH_3)_3$

2d.

O
(H3C)2C—CH2
O
⊖ CH2—S—(CH3)2
⊕

O
⊖ CH2—S—(CH3)2
O ⊕
(H3C)2C—CH2

O—CH2
(H3C)2C—CH2 + (H3C)2S=O

2e.

CH3
O
H3C
O
⊖ CH2—S(CH3)2
⊕

CH3 O⊖
O
—CH2—S(CH3)2
⊕
H3C

CH3 O⊖
O
C—CH2—S(CH3)2
⊕
H3C CH2

CH3 O
O
C—CH2—S(CH3)2
⊕
H3C CH2
⊖

(CH3)2S=O +
CH3 O
H3C

2f. Br C=O CH3 + ⊖CH2—S(CH3)2 (O, ⊕) ⟶ ⊖CH2 C=O Br ⟶ O + ⊖CH2—S(CH3)2 ⟶ O⊖ CH2—S(CH3)2 ⟶ O

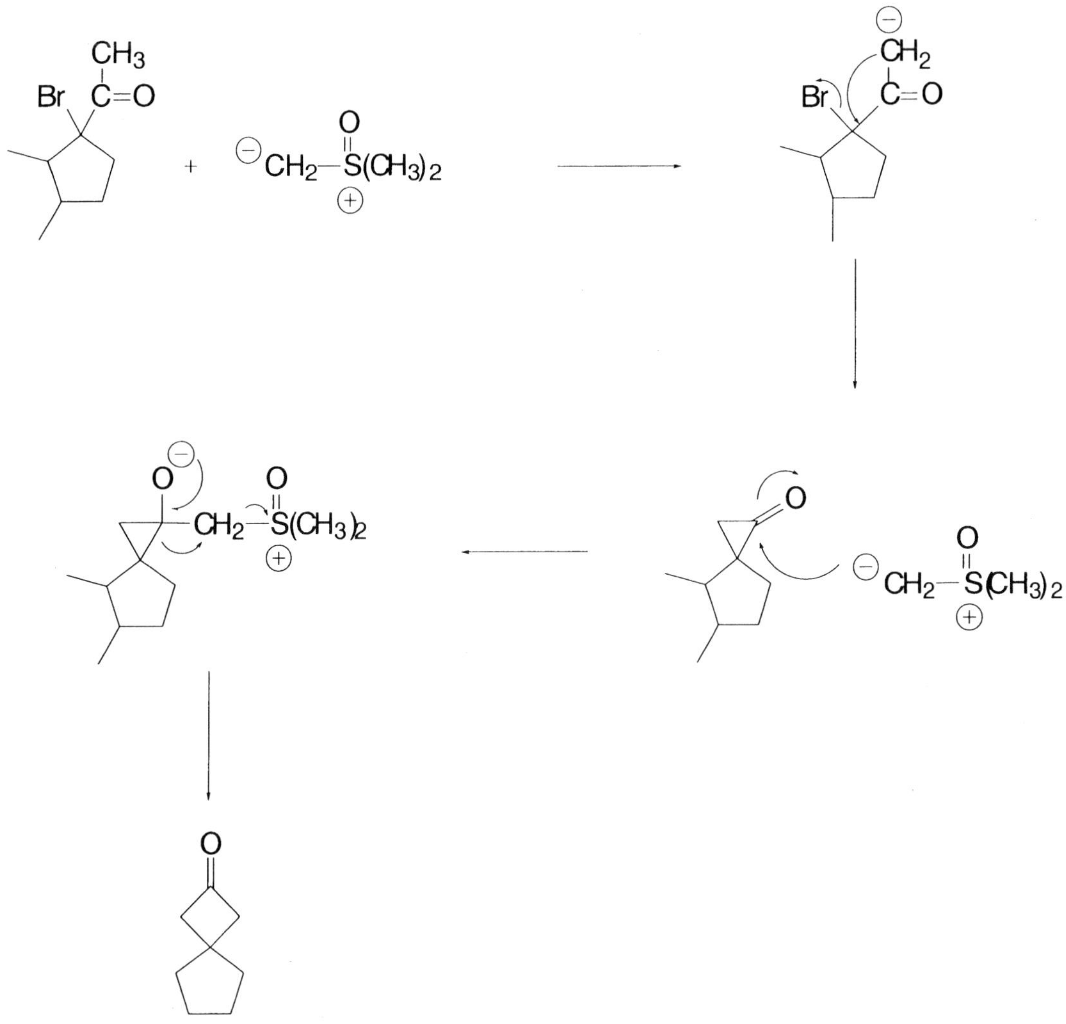

2g. $(CH_3)_2S$ + :CH_2 ⟶ $(CH_3)_2\overset{\oplus}{S}-\overset{\ominus}{CH_2}$ + O ⟶ O⊖ $H_2C-\overset{\oplus}{S}(CH_3)_2$ ⟶ O CH_2 + $(CH_3)_2S$

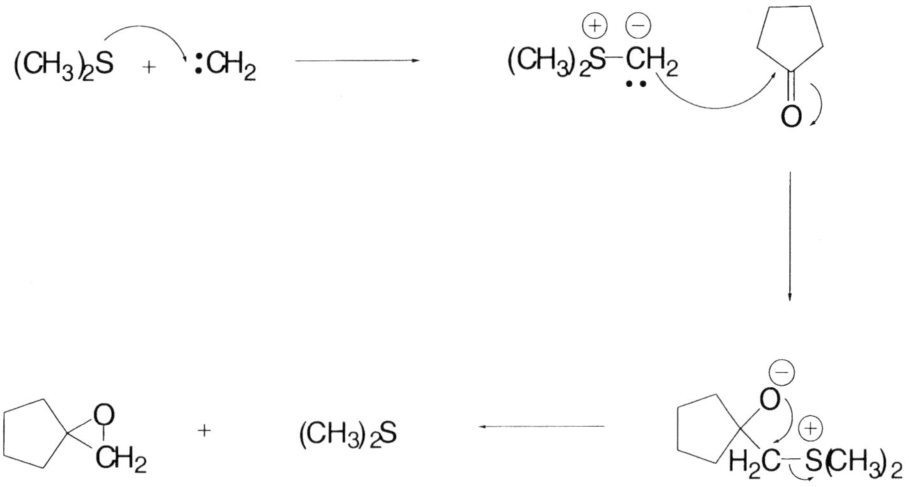

2h.

$(C_6H_5)_3P\colon$ + $Br{-}CBr_3$ ⟶ $(C_6H_5)_3\overset{+}{P}{-}Br$ + $^{-}CBr_3$ ⟶ $(C_6H_5)_3\overset{+}{P}{-}CBr_2Br$ + Br^{-}

$(C_6H_5)_3\overset{+}{P}{-}CBr_2{-}Br$ + $(C_6H_5)_3P$ ⟶ $(C_6H_5)_3P{=}CBr_2$

$(C_6H_5)_3P{=}CBr_2$ + C_6H_5CHO ⟶ $C_6H_5CH{=}CBr_2$ (Wittig Reaction)

2i.

2,6-dibromo-4-bromo-4-methylcyclohexa-2,5-dienone + $\colon P(OCH_3)_3$ ⟶ 2,6-dibromo-4-methylphenoxide (O^{-}) + $Br{-}\overset{+}{P}(OCH_3)_3$

⟶ $ArO{-}\overset{+}{P}(OCH_3)_2{-}O{-}CH_3$ + Br^{-} (Ar = 2,6-dibromo-4-methylphenyl)

⟶ $ArO{-}P(=O)(OCH_3)_2$ + CH_3Br

2j.

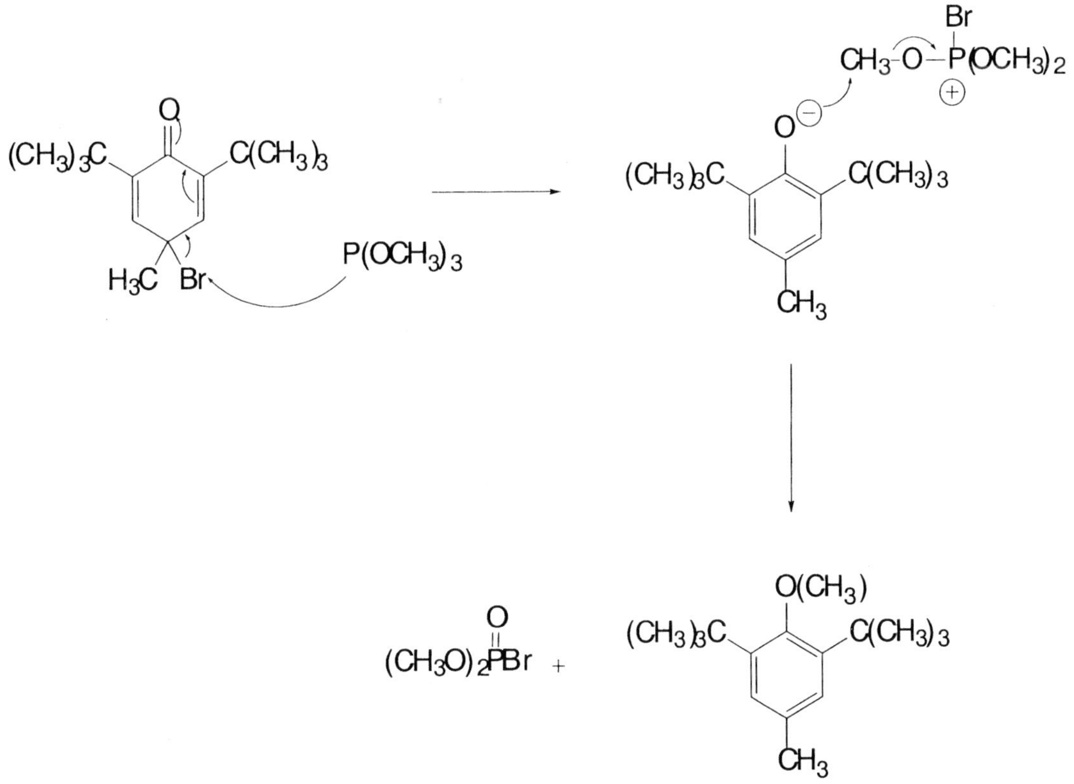

$(CH_3)_3C$ $C(CH_3)_3$ O H_3C Br $P(OCH_3)_3$ → $CH_3-O-\overset{Br}{P}(OCH_3)_2$ ⊕ $O^{\ominus}$ $(CH_3)_3C$ $C(CH_3)_3$ CH_3

↓

$(CH_3O)_2\overset{O}{\overset{\|}{P}}Br$ + $O(CH_3)$ $(CH_3)_3C$ $C(CH_3)_3$ CH_3

2k.

$\overset{O}{\overset{\|}{C}}OC_2H_5$ $^{\ominus}CH_2-\overset{O}{\overset{\|}{S}}(CH_3)_2$ ⊕ → $\overset{O}{\overset{\|}{C}}OC_2H_5$ ⊖ H–CH_2 $CH_2-\overset{\oplus}{S}-CH_3$ O

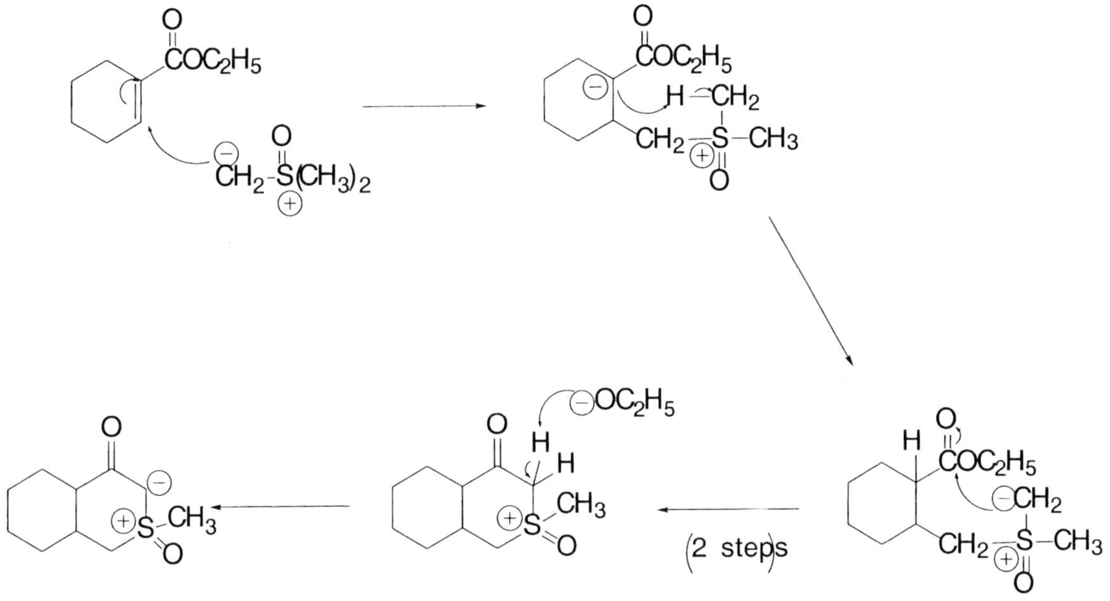

H $\overset{O}{\overset{\|}{C}}OC_2H_5$ $^{\ominus}CH_2$ $CH_2-\overset{\oplus}{S}-CH_3$ O ← (2 steps) O H H $^{\ominus}OC_2H_5$ ⊕S CH_3 O ← O ⊖ ⊕S CH_3 O